On Systems Analysis:
An Essay Concerning the
Limitations of Some
Mathematical Methods in
the Social, Political,
and Biological Sciences

On Systems Analysis:
An Essay Concerning the
Limitations of Some
Mathematical Methods in
the Social, Political,
and Biological Sciences

David Berlinski

The MIT Press
Cambridge, Massachusetts,
and London, England

To Toby

SCI
QA
402
B48

This book was set in I.B.M. Century by Jay's Publishers Services, Inc. and printed and bound by The Colonial Press Inc. in the United States of America.

Second printing, 1977

Library of Congress Cataloging in Publication Data

Berlinski, David J
    On systems analysis.

    Includes index.
    1. System analysis. 2. System theory.
I. Title.
QA402.B48   003   76-13444
ISBN 0-262-02120-X

Tout est complexe dans la nature. . .

René Thom

Contents

Preface

In the spring of 1970, Jules Cohn and I found ourselves employed by McKinsey and Co., Inc. It was a time when "systems analysis" was yet a purely honorific term. Professor Cohn became the guest editor for the fall issue of the *Urban Affairs Quarterly*, and knowing of my skeptical strictures regarding the systems sciences, he asked me to contribute to his issue. I did and the essay that resulted, parts of which appear again in the first chapter of this book, made its way through the agency of Aaron Wildavsky to Philip Siegelman, the review editor of *The American Political Science Review*. Professor Siegelman encouraged me to prepare an article on much the same topic for the *Review*; I accepted his invitation, but it soon became clear that the material I had on hand would prove too copious for a journal. Instead I took the advice of an anonymous reader, who assured me that in my review essay I had "the core of a good and useful book," and restored to legitimacy chunks of criticism that I had suppressed in the interests of space. Thus fattened, the review essay that I originally prepared for *The American Political Science Review* became a monograph, the monograph finally this book-length essay.

Friends who have watched the work grow have sometimes asked why I spend so much time denouncing as worthless theories that *are* worthless. To this question I have no reasonable answer save the one that H. L. Mencken gave: for the same reason that men go to zoos.

Some critics who have read bits and portions of my book have come away convinced that I am opposed to the use of mathematics in the social, political, or biological sciences. Nothing could be further from the truth. I am obliged to emphasize this point so as to be spared the friendship of my enemies. It is the use of mathematical methods for largely ceremonial reasons that I deplore and denounce as pernicious. The ability to see in the shifting and confused realm of appearances the outlines of a stable system of abstract mathematical objects is, quite simply, the very essence of intellectual insight.

Ultimately, I see my essay as a fragment of a larger and philosophically more stimulating mass dealing with the question of the extent to which minds constituted roughly as ours are constituted may hope to achieve an adequate understanding of social, political, and biological life. There is no obvious reason why the interiors of certain distant stars should be the subject of deep physical theories while the gross features of ordinary experience remain, by the standards of theoretical physics or pure mathematics, inaccessible to sophisticated speculation. Systems analysis I take to be a bouncy and outrageous attempt to pretend that such speculation, and the understanding that it sometimes prompts, is in fact within our grasp; to show that this is not so is not a very difficult task, but it is one that ought to be performed.

My friend Professor Murray Lieb has carefully read this essay, and I am grateful to him for the corrections that he suggested. Daniel Gallin read earlier sections of the manuscript; I am indebted to him in a larger sense for affording me a model of mathematical tact and probity.

New York, 1976

Introduction

Systems analysis, like Ethnic Studies, admits of a division into theoretical and applied wings. To the former goes the elaboration of the
concept of a system and all its laws; by my count, this takes in general
systems theory, mathematical systems theory, which has a journal to
itself, cybernetics (the Fair One), information theory, control theory,
optimal control theory, stochastic control theory, even set theory,
automata theory, and the theory of ordinary differential equations; and
to the latter, the invocation of those resuscitative arts—program bud-
geting, cost-benefit analysis, and the like—that systems analysts assure us
bring to the operations of government a kind of sanctified effectiveness
impossible to achieve under conditions of routine political piggishness.

One cannot discuss everything. This is the abiding sadness of the
reviewer. In what follows I slight the resuscitative arts. My own concerns are strictly theoretical. I stick throughout to a three-headed segregation of topics, with the first third of the book devoted to general
systems theory, the second to dynamical systems, and the third to
mathematical systems theory.

1 General Systems Theory

We postulate a new discipline called *General System Theory* . . . a logico-mathematical field whose task is the formulation and derivation of those general principles that are applicable to "systems" in general.

Ludwig von Bertalanffy [1]

The Laws of Systems:
Under the Aspect of
von Bertalanffy

Considering how much mathematics there is, it is remarkable
how little of it is of any use.

Old Saying

General systems theory is nothing if not ambitious. In the opening
chapter of his text, Professor von Bertalanffy claims information theory, set theory, graph theory, game theory, decision theory, the theory
of automata, and cybernetics as parts of GST. The sense of motley
abandon is reinforced both by a reading of *General System Theory*
and by thumbing through the yearbooks of the Society for General
Systems Research.[2] The latter especially contain papers on an aston-
ishingly diverse number of subjects. But Professor von Bertalanffy, by
his own argument, does not preside over a zoo; the discipline has a purpose:

Its subject matter is the formulation and derivation of those principles
which are valid for "systems" in general.
    [Its] meaning . . . can be circumscribed as follows. . . . [We] can ask
for principles applying to systems in general, irrespective of whether
they are of physical, biological or sociological nature. If we pose this
question and conveniently define the concept of system, we find that
models, principles, and laws exist which apply to generalized systems
irrespective of their particular kind, elements, and the "forces" involved.
    A consequence of the existence of general system properties is the
appearance of structural similarities or isomorphisms in different fields.
There are correspondences in the principles that govern the behavior of
entities that are, intrinsically, widely different.[3]

A discipline so gigantically construed suggests at least three questions:

---

[1] Ludwig von Bertalanffy, "The History and Status of General Systems Theory," in
George J. Klir, ed., *Trends in General Systems Theory* (New York: Wiley-Interscience, 1972), p. 26.

[2] The basic text in the analysis of GST is, of course, Ludwig von Bertalanffy, *General System Theory* (New York: George Braziller, 1968). The book includes an
extensive bibliography. The Society for General Systems Research publishes a
yearbook, *General Systems*, and a quarterly journal, *General Systems Journal*.
Yearbook articles are often quite interesting but rarely, so far as I can see, have
anything to do with GST. Ervin Laszlo has edited an absolutely hilarious series on
GST for the International Library of Systems Theory and Philosophy, published by
George Braziller. GST has exerted an irresistible attraction on the academic crank:
books like V. L. Parsegian's *This Cybernetic World of Men, Machines and Earth
Systems* (New York: Doubleday, 1972), Milton D. Rubin's *Man in Systems* (New
York: Gordon and Breach, 1971), and Arthur Iberall's *Toward a General Science of
Viable Systems* (New York: McGraw-Hill, 1972) appear with metronomic regularity.

[3] Von Bertalanffy, *General System Theory*, pp. 32–33.

(1) What are systems? (2) What are some of their laws or properties? (3) What interesting isomorphisms between systems does GST illuminate?

A. D. Hall and R. E. Fagen, in a paper entitled "Definition of System," have thoughtfully addressed first matters first:
- A system is "a set of objects together with relationships between the objects and between their attributes."
- Objects are simply "the parts or components of the system."
- Attributes are "properties of objects."
- Relationships are those things that "tie the system together."[4]

Just so.

These citations, plucked as they are from a more discursive article, illuminate at low wattages. Hall and Fagen have suggested something that mathematicians would recognize as an elephantine approximation to the notion of a model. The term "model," of course, enjoys almost the same high status as the term "system" and, one might add, is generally employed with the same deft precision.[5] Greater clarity is possible if we hew to the standard notion of a model employed in modern model theory or semantics.

What is called for formally is an explicit choice of language—L, say—whose vocabulary usually comprises an infinite list of individual variables, standard sentential connectives, quantifiers, and at most denumerably many predicate variables of various finite ranks. (A predicate's "rank" betokens the number of its argument places: a one-place predicate denotes a property; a two-place predicate, a relation; and so on up to the various n-place predicate symbols.) A model can then be defined as an ordered pair $M = <D, F>$, where D is a nonempty set (the domain of M) and F is a function assigning to the predicate variables of L relations of corresponding rank on D. The values of F are called the *relations* of M. Given this much, it is possible to define important semantic relationships between L and M. An assignment α is a function that maps the individual variables of L onto individuals in the domain D. The notion

α satisfies the formula $S(x, \ldots)$ in M

[4] A. D. Hall and R. E. Fagen, "Definition of System," in N. Buckley, ed., *Modern Systems Research for the Behavioral Scientist* (Chicago: Aldine Press, 1968).
[5] See Patrick Suppes, "A Comparison of the Meaning and Uses of Models in Mathematics and the Empirical Sciences," in the collection of his papers, *Studies in the Methodology and Foundations of Science* (New York: Humanities Press, 1969), for a workmanlike discussion of the various senses connoted by the term "model."

can then be defined by recursion on the length of the formula $S(x, \ldots)$. The intuitive idea behind the definition is that an assignment α satisfies a formula if the formula holds in M, when its predicate variables are interpreted by F, its free individual variables are interpreted by α, and its bound variables range over the domain D. To say that the definition is recursive (rather than explicit) means merely that satisfaction is defined for formulas of a given length in terms of a definition of satisfaction already voiced for formulas of shorter length—this for all formulas save the shortest, for which satisfaction is defined explicitly.

A sentence, which is a formula with no free individual variables, will be satisfied either by every assignment in M or else by none. In the first case the sentence is true in M, otherwise false; the theory T of the model M is identified simply as the *whole* of the set of sentences satisfied in M.

The usage thus employed sharply distinguishes between a model, which is a set-theoretical entity, and the language that is used to talk about the model, which, of course, is not. This distinction is frequently blurred in discussions of systems analysis and often, in what amounts to a systematic confusion of use and mention, the passage from model to theory and back to model again is made with oleaginous ease.

So systems are models, at least on this unsolicited reconstrual of GST. That takes care of (1). What of (2)? The notion of a law is unclear, but for purposes at hand, noting simply that the laws of L displace a certain space within its *truths* should suffice. In view of the proposed generality of GST, then, it would seem natural to require that a sentence be true in all systems if it is to be among the truths of GST. But the *logical* truths are true in all systems, so the laws of GST must be a trivial subset of the laws of logic.[6]

This is a bad beginning. Professor von Bertalanffy, it is true, does not hold consistently to the line that systems are models.[7] His arguments are informal and involve nothing like an unambiguous commitment to systems taken in a particular sense. Nor need they. But many of his examples are distinctly unwholesome, and there is a strong tug in the explanations that accompany them back toward the triviality that the laws of systems make up a subset simply of the laws of logic. The early

---

[6] See D. Berlinski and D. Gallin, "Quine's Definition of Logical Truth," *Nous*, Vol. 3 (1969), pp. 1–7. Carl Hempel makes what I take to be a similar argument in his essay "General Systems Theory and the Unity of Science," *Human Biology*, Vol. 23 (December 1951), pp. 313–322.

[7] But see von Bertalanffy, *General System Theory*, p. 95.

chapters stick to systems taken as a string of simultaneous differential equations:

$$\frac{dQ_1}{dt} = f_1(Q_1, Q_2, \ldots, Q_n),$$

$$\frac{dQ_2}{dt} = f_2(Q_1, Q_2, \ldots, Q_n),$$

$$\tag{1.1}$$

$$\cdot$$
$$\cdot$$
$$\cdot$$

$$\frac{dQ_n}{dt} = f_n(Q_1, Q_2, \ldots, Q_n).$$

This indicates, of course, an atherosclerotic narrowing of GST's overall program: laws true in just such systems will not be true in *all* systems. Still, the scientist, sustained by GST, can by surveying these systems discover "certain laws of nature . . . not only on the basis of experience, but also in a purely formal way,"[8] and it is this claim that remains redemptively distinctive to GST. I am not perfectly sure what it means, but I would guess that it involves the avowal ultimately that the laws of nature, having been discovered in whatever fashion—by intuition, hunches, or the apparatus of statistical decision theory—can also be formally derived by setting them in the context of some more embracing axiomatic theory based on still other laws of nature, much as Kepler's laws can be formally derived from Newton's laws or Newton's laws from postulates such as Maupertuis's principle of least action.

I myself take the position that reregistration of scientific theories in axiomatic form reflects a barren architectural passion—down Hilbert, down good dog—and while I admire the man who takes formal methods to subjects precise within the limits of ordinary mathematical practice, my enthusiasm for his zeal is corrupted by my indifference toward his aims. In fact, the axiomatic von Bertalanffy gets only the briefest play. "We can ask," he writes instead,

for *principles applying to systems in general,* irrespective of whether they are of physical, biological or sociological nature. If we pose this question and conveniently *define the concept of a system,* we find that

---

[8] Ibid., pp. 62-63.

models, principles, and laws exist which apply to *generalized systems* irrespective of their particular kind.[9]

By a generalized system I take it that von Bertalanffy means merely a set of systems defined narrowly enough to prevent GST from caving in toward the purely logical truths. To the theoretician goes the twofold task of specifying such systems and then formally coaxing from the definitions those laws and principles that hold good in *each* of the models that count as instances of the generalized system. The string (1.1), I take it, is an example of just such a system.

It is, at any rate, (1.1) from which laws of nature are to issue. Actually, von Bertalanffy's argument, such that it is, involves rather an immediate compression of elements figuring in (1.1), so that what one works with is not a system of simultaneous differential equations but the single equation

$$\frac{dQ}{dt} = f(Q); \tag{1.2}$$

$f(Q)$ is promptly expanded in a Taylor series

$$\frac{dQ}{dt} = a_1 Q + a_{11} Q^2 + \ldots \tag{1.3}$$

which is then summarily truncated at its first term:

$$\frac{dQ}{dt} = a_1 Q. \tag{1.4}$$

This, von Bertalanffy holds, corresponds somehow to "the simplest possibility." The equation that results is, of course, an old standby whose solution is exponential in t:

$$Q = Q_0 e^{a_1 t}. \tag{1.5}$$

Von Bertalanffy calls (1.5) a *law of nature*: the whole alarmingly rapid passage from (1.2) to points that follow

[illustrates] a point of interest for the present consideration, namely the fact that certain laws of nature can be arrived at not only on the basis of experience, but also in a purely formal way. The equations discussed signify no more than that the rather general system of equa-

---

[9] Ibid., p. 33; emphasis added.

tion [has been developed] into a Taylor series and suitable conditions have been applied. In this sense such laws are "*a priori*" independent from their physical, chemical, biological, sociological, etc., interpretation. In other words, this shows the existence of a general system theory which deals with formal characteristics of systems, concrete facts appearing as their special applications by defining variables and parameters. In still other terms, such examples show a formal uniformity of nature.[10]

In the context of GST this argument is representative: other examples deal with *competition, wholeness, mechanization,* and *finality*, system-wide laws appearing as solutions to an appropriate differential equation.[11]

Equation (1.4) does have modest usefulness in the description of *uniform growth*—a continuously copulating clutch of rabbits, for example. Under suitable idealization it will also describe the behavior of populations that grow by a fixed integer—ordinary biological populations, bank balances, systems analysts. The metric volume of confusion in this passage is nonetheless considerable. The first sentence suggests somehow that being *based in experience* and being *formally derivable* are alternative but symmetrical procedures whereby a sentence may be counted as a law of nature. This collapses the distinction between inductive and deductive experience. What von Bertalanffy means, no doubt, is that from suitably general definitions certain statements turn up under the influence of GST that *turn out to be* laws of nature—when, for example, they are widely confirmed.

With distinctions thus fixed, scan again equations (1.2) to (1.5). Taylor's theorem and its corollary, which sanction the expansion of (1.2) into (1.3), are, of course, theorems of analysis; similarly the conditional whose antecedent is (1.4) and whose consequent is (1.5). That these sentences turn out to be true in models other than the primary models of analysis is hardly surprising: if a complex theory T, developed to discuss population growth, invokes the powers of the calculus, then models of T must also be rich enough to satisfy portions of analysis thus developed.

But, my more clamorous readers will insist, von Bertalanffy is not arguing merely that the theorems of the calculus turn out to be true wherever they *are* true: when properly reinterpreted, some hold of a variety of nonmathematical entities as well—(1.5) is an example. It is

[10] Ibid., pp. 62–63.

[11] These arguments appear in Chapter 3 of *General System Theory*; the rest of the book wanders over a variety of biological topics; von Bertalanffy was a talented speculative biologist and his stuff on the principles of growth, allometry, evolution, and so forth makes for interesting reading.

wonderful that one statement holds of physical entities when we measure population growth, of bank balances when we compound interest, and of bacteria when we cultivate *Staphylococci pyogenes var. aureus.*

Notice, however, that though the entailment of (1.5) by (1.4) ranks as a theorem of the calculus, or would were relevant portions formalized, (1.5) does not *thereby* count as a truth of genetics, for example, or the theory of population growth. In fact, there are obvious models in which it turns out false, even though all but its parameters retain their standard interpretations. To register it as a law of a specific field on the strength of its heritage in the calculus, one must confirm it to be true in its reregistered interpretation by showing that it is (1.4) that accurately describes the facts and not some other equation.

No doubt it is still wonderful that (1.5) and its crowd happen to be confirmed in so many different domains. So long as my readers hew to the cited distinctions and urge no truths electrifyingly discovered by unaided reason, I find no fault and plan to content myself with the murmuring of a few academic "hear, hear's." (But if extended reinterpretability is what is wanted, why not count as truths—and hence as laws—of GST only statements which like the laws of logic are satisfied in every model or system?)

The grand aims of GST, however, call for something more ambitious than the uncovering of a formula that is widely confirmed: what needs to be shown is that (1.4), together with its solution, is true in *all* systems whose defining equations may plausibly be taken as instances of (1.1). But here systems theorists confront The Crunch. The basic strategy in the course from (1.1) to (1.5)—from the definition of systemhood to the first (and so far the only) law of systems—involves an *argument by accretion* trading on the steady if surreptitious accumulation of assumptions to grease the way from an initial premise to a settled conclusion which it does not otherwise imply. In the present case, accreted premises are set out at (1.2), (1.3), and (1.4); of this process von Bertalanffy writes that "in every hypothetico-deductive system . . . we have to introduce *special conditions* according to experience in order to apply it to concrete phenomena." A good example, he adds optimistically, "is the theory of populations. . . . Starting from a general and purely formal set of equations, and gradually introducing more and more specific conditions, laws for the growth of populations consisting of one or more species . . . can be derived."[12] So one really has an avowed argument by accretion, a polemicist's dream.

---

[12] Ludwig von Bertalanffy, "General Systems Theory: A New Approach to the Unity of Science," *Human Biology*, Vol. 23 (December 1951), p. 339; emphasis added.

The Crunch works via a destructive dilemma. Taking (1.1) as comprising a class of general systems and (1.5) as the law derived by purely "formal" considerations, we split the field of possibilities lengthwise: if the systems are general, the derivation is incorrect; if the derivation is correct, the systems are not general. Of course, (1.5) is simply the solution to (1.4)—an algebraic statement to the effect that the unknown function Q at (1.4) is exponential. But not all solutions to equations that constitute instances to (1.1) are exponential; for example,

$$\frac{dQ}{dt} = Q^2 \tag{1.6}$$

is of the form (1.2), but its solution is

$$Q = \frac{1}{c - t}, \tag{1.7}$$

where c is an arbitrary constant. This is The Crunch worked from right to left. What we must assume, in order to come up with systems to which (1.5) expresses an appropriate differential solution, is expressed at (1.2), (1.3), and (1.4); there the natural sense of systemhood is systematically contracted so that what counts as a system is tacitly defined *not by (1.1) but by (1.4) instead*. But (1.4) is an equation of modest scope describing systems in uniform growth: there are other systems of growth and other differential equations to describe them (e.g., the Malthusian equation $du/dt = Ku(U - u)$, in which the *unit rate* $K(U - u)$ decreases as $u \to U$). This is The Crunch worked from left to right.[13]

The idea that very particular empirical principles can somehow be eked out of overwhelmingly general definitions betrays an almost Oriental passion for the union of opposites. Count *anything* as a system and your theory will include just the logical truths—statements that are all form and no content: if P then P or Q. Something more significant

[13] The redoubtable Daniel Gallin writes me that

as far as the philosophical conclusions Bertalanffy draws from this rather uninspired example, he has not discovered any "truth holding in all systems" at all, but simply solved (incorrectly, to be sure) a simple differential equation. Laws of nature do not come into the picture anywhere. If, in some particular application, the differential equation (1.4) holds—or holds in the idealized case—then the function in question is exponential. That's all there is to it. He has no more arrived at a "law of nature" this way than if one wrote down $3 + Q = 2Q$, proceeded to derive $Q = 3$, and then announced the discovery of a law of nature in a "purely formal way." The whole example is pure nonsense.

A harsh conclusion, no doubt, but one to which a man might be pushed. Von Bertalanffy himself calls it to mind on p. 35 of *General System Theory*.

can be achieved only if the structures in which a set of sentences is satisfied are correspondingly augmented, and this inevitably involves a decisive reduction in the number of algebraic entities that will satisfy the sentences of a given theory. Thus the axioms and theorems of group theory, when set against the trivial panorama afforded by the purely logical truths, represent a gain in content but a contraction in scope, since the principles that are true in *all* groups are not true in *all* systems.

These observations spell the end, so far as I can see, of GST as a science specialized somehow in the extraction of general laws from curiously arbitrary definitions. The point needs to be stressed, I suppose, that what ails GST, at least in the fashion that von Bertalanffy saw it, is *not* the harmless enough insistence that science amounts to the unraveling of deductive consequences from suitably framed definitions: all sophisticated science involves the creation of mathematical structures as representatives of reality. (This is a concession that may be made without confusing theory with reality. The status of a given principle as a law of nature depends chiefly on its success in characterizing reality and not on its role in a mathematical structure.) But GST involves the impulse to carry the dramas of mathematical construction to a point at which the entities that result are so painfully swollen and senselessly general that without the surreptitious introduction of a series of corset-like contracting definitions *nothing* of any interest follows deductively. Mathematical structures that are very general cannot be rich.

There yet remains, perhaps, a purely *taxonomical* role for GST, one that makes something of the general character of the differential equations at (1.1) without trading on bogus deductions. If *any* set of equations of the form described by (1.1) counts as an example of a system, the systems analyst having them on hand has in the nature of things a complete catalog of all possible laws of nature, at least insofar as the laws involved are (1.1)-like. It remains, then, only for him to effect a winnowing among the infinitely many functional relationships instanced at (1.1) in order to come up with just the laws that might hold of physical systems. This, I gather, is an interpretation that Hempel finds persuasive; he writes that "the task of GST would presumably consist in a study of various mathematical forms of interdependence between the constituent parts of a physical object, in regard to one or more quantitative characteristics."[14] A truly general theory of systems, he adds reasonably, "would have to investigate all *possible* types of functional interrelationships"—a mathematical investigation, one would think, of Brobdingnagian proportions. But while it is true that such a

[14] Hempel, "GST and the Unity of Science," p. 314.

catalog of all possible forms of functional interrelationship would sweep the laws of nature as well as everything else, having *every* differential equation one still does not know *which* might be the laws of nature. Here, as elsewhere, Hempel is especially acute: "From von Bertalanffy's writings," he remarks,

I gather that he would hardly want to consider every conceivable type of functional relationship as representing some potential form of synholistic interdependence, and that he would rather consider systems laws as some proper subclass of all possible functional relationships. However, he has not offered any general criterion to characterize the intended subclass, nor is such a criterion obvious.[15]

Thus, even in its more modest role as a purveyor of possible systems of functional dependencies, obvious obstacles swarm up out of the fog to bedevil GST—evidence, I should think, of fundamental obscurities in its overall conception.

[15] Ibid.

The Properties of Systems:	Ours is a complex world.
Under the Aspect of Laszlo	E. Laszlo[16]

No doubt. Some succor is afforded by the "disciplinary matrix of General Systems Theory," an object of splendid theoretical merits, consecrated to *holism* in ontology, *integration* in epistemology, *unity* in philosophy, and *humanism* in science.[17] Professor Laszlo's own efforts extend chiefly to the *natural systems* and those invariances that apply to phenomena of organized complexity throughout the microhierarchy.[18]

Since the systems sciences are "mathematical in form,"[19] Laszlo commences each section of the chapter in *Introduction to Systems Philosophy* that outlines a "general theory of systems" with a tidy equation. But what technical material there is, it turns out, is entirely

[16] Ervin Laszlo, *The Systems View of the World* (New York: George Braziller, 1972), p. 13.

[17] Ervin Laszlo, ed., *The Relevance of General Systems Theory* (New York: George Braziller, 1972), pp. 4-6.

[18] Ervin Laszlo, *Introduction to Systems Philosophy* (New York: Harper & Row, 1972), p. 35. This is the volume for those wishing to mush on "Towards a New Paradigm of Contemporary Thought."

[19] Laszlo, *The Relevance of General Systems Theory*, p. 9. The intoxication with mathematical methods is, of course, general throughout the social sciences. An announcement from the Princeton University Press brings news of a "major new series to encourage the application of mathematical methods to historical analysis"; the first title on the list, W. O. Aydelotte, A. G. Bogue, and R. W. Fogel's *The Dimensions of Quantitative Research in History*, looks for all the world like a volume authored by the Bourbaki. See also H. R. Alker, K. W. Deutsch, and A. H. Stoetzel, eds., *Mathematical Approaches to Politics* (New York: Elsevier, 1973). Theories that make use of mathematical methods need not involve the attribution of significant mathematical structure. Regression analysis is a case in point; theories of price equilibria invoking commodity spaces mark the contrast. In the social sciences only economics is committed to a wide range of interesting and suitably abstract mathematical objects. See, for example, E. Dierker's very interesting monograph, *Topological Methods in Walrasian Economics* (New York: Springer-Verlag, 1974). But many analyses in political science exude an attractive incompleteness that might be made good by reference to precisely fixed mathematical objects. David Gale's "On the Theory of Interest" (*American Mathematical Monthly*, Vol. 80, No. 8) has some interesting remarks on related topics. See also the editorial comment in the June 1974 issue of the *American Political Science Review* (Vol. 118, No. 2). There is now much concern among professional mathematicians—I except the statisticians—for the superb inutility of their subject. These are hard times: not a month now passes without some or another mathematician musing in the mathematical journals whether a life spent by the lamp searching for an ever larger inaccessible cardinal may not prove to be an economic disaster, whatever its aesthetic attractions.

ceremonial: "I shall not burden the reader," Professor Laszlo explains in an early footnote, "with the mathematics."

It is a pity.[20] The missing mathematics might have shown that natural systems are "other than the simple sum of the properties and functions of their parts."[21] As it is, Laszlo argues only that the natural systems, as opposed to heaps, are "nonsummative," possessing characteristics "not possessed by [their] parts singly."[22] But so do heaps since only the whole of a heap of systems theorists, for example, is the *whole* of that heap. To my way of thinking this suggests that heaps are nonsummative; but Laszlo has it the other way around, arguing that heapwise "it is sufficient to sum the properties of the parts in order to obtain the properties of the whole."[23]

[20] Perhaps it is just as well: the example that Laszlo chooses to illustrate nonsummative complexity is just a system of ordinary differential equations; here, Laszlo claims, "change of any measure Q_i is a function of all Q's and conversely" (*Introduction to Systems Philosophy*, p. 37). By "measures" Laszlo no doubt means functions.

[21] "By a natural system," Laszlo writes on p. 30 of his *Introduction to Systems Philosophy*, "I understand that which is sometimes referred to as a *concrete* system; i.e., a 'nonrandom accumulation of matter-energy, in a region of physical space-time, which is nonrandomly organized into coacting interrelated subsystems or components.' " The quotation is from James G. Miller, "Living Systems: Basic Concepts," *Behavioral Science*, Vol. 10, pp. 193-237, and, as far as I can judge, totally without sense.

[22] Laszlo, *Introduction to Systems Philosophy*, p. 36.

[23] Ibid., p. 37. The concern for the maxim that the whole is more than the sum of its parts extends to sociology, a discipline occasionally in the service of GST: what follows is from Walter Buckley's *Sociology and Modern Systems Theory* (Englewood Cliffs, N.J.: Prentice-Hall, 1967), p. 42:

When we say that "the whole is more than the sum of its parts," the meaning becomes unambiguous and loses its mystery: the *more than* points to the fact of organization which imparts to the aggregate characteristics that are not only *different* from, but often *not found in* the components alone; and the "sum of the parts" must be taken to mean, not their numerical addition, but their unorganized aggregation.

The essential point, I would argue, is that an object, heap or whole, is complex or simple, organized or unorganized, not in the fullness of any absolute ontological sense, but with respect to some or another specified theory. Thus pairs of objects rank in complexity only *within* some given theoretical frame: an object rendered heaplike from the perspective of theory T may decompose into a tight network of regular relations from the perspective of theory T'. Ernest Nagel discusses this issue at some length in *The Structure of Science* (New York: Harcourt, Brace & World, 1961), pp. 380-398.

That a set of objects organized by some joint relation or relations possesses characteristics that are not only "*different* from, but often *not found in* the components alone" follows, of course, from the fact that these characteristics are different from any found in the components; this, in turn, simply means that relations are

Whatever the heap-whole distinction, natural systems fall under the purview of *systems cybernetics*, a science devoted to positive and negative feedback and uniquely specialized in the illumination of those *invariants* that characterize the natural systems. Such systems have, for example, powers and properties of adaptive self-stabilization:

Ordered wholes are always characterized by the presence of fixed forces, excluding the randomness prevailing at the state of thermodynamic equilibrium. Thus ordered wholes, by virtue of their characteristics, are self-stabilizing in, or around, steady states.[24]

But this is an appraisal of indecently encouraging vagueness, given Laszlo's belief that atoms, molecules, crystals, cells, viruses, organisms, ecologies, and societies are all natural systems. Indelicate, too, is the transference of purely thermodynamic considerations to the Federal Housing Administration or the principal industrial sectors of the Baltic states, especially since the *evidence* for adaptive self-stabilization is derived, on Laszlo's account, chiefly from Le Châtelier's principle, a maxim from the elementary theory of chemical equilibria summarizing Robin's and van't Hoff's equilibrium laws and having, in any case, nothing whatsoever to do with adaptive self-stabilization or the properties of ordered wholes.[25]

Adaptive self-stabilization is made no clearer by the invocation of *cybernetic stability*. "Let us focus," Laszlo writes, quoting Paul Weiss,

---

ineliminable in favor of monadic predicates, a true enough hypothesis so far as formal logic goes, but one that, again, needs restriction to specific theories. Thus, on some versions of arithmetic, the relation "*a* is *between c* and *d*" might be defined in terms of the relations "less than" and "greater than," with these defined in terms of primitives such as membership. The primitives constitute the theory's ineliminable relationships. The general idea that relations are ineliminable *überhaupt* is, I suppose, what Charles Peirce meant when he spoke of the ineliminability of thirdness.

[24] Laszlo, *Introduction to Systems Philosophy*, p. 40.

[25] Paul Samuelson remarks, in his Nobel Lecture, that

there is nothing more pathetic than to have an economist or a retired engineer try to force analogies between the concepts of physics and the concepts of economics. How many dreary papers I have had to referee in which the author is looking for something that corresponds to entropy or to one or another form of energy. Nonsensical laws, such as the law of conservation of purchasing power, represent spurious social science imitations of the important physical law of the conservation of energy; and when an economist makes reference to a Heisenberg principle of indeterminacy in the social world, at best this must be regarded as a figure of speech or a play on words.

Strong words these, especially since they are made in the context of a discussion of Le Châtelier's principle and its relationship to economics ("Maximum Principles in Analytical Economics," *Science*, Vol. 172 [September 10, 1971], p. 994).

on any fractional part (A) of a complex suspected of having systemic properties, and measure all possible excursions and other fluctuations about the mean in the physical and chemical parameters of that fraction over a given period of time. Let us designate the cumulative record of those deviations as the variance (v_a) of part A. Let us furthermore carry out the same procedure for as many parts of the system as we can identify, and establish their variances v_b, v_c, v_d, . . . , v_n. Let us similarly measure as many features of the total complex (S) as we can identify and determine their variance (V_s). Then the complex is a system if the variance of the features of the whole collective is significantly less than the sum of the variances of its constituents; or written as a formula,

$$V_s \ll \sum(v_a + v_b + \ldots + v_n).^{26}$$

I am not sure what variance comes to in this context: I count it a synonym for *change*; systems but not heaps change less than the sum of the changes of their parts, and it is this property that further separates the adaptively self-stabilizing systems from the heaps merely. But the characterization of cybernetic stability that Weiss offers and Laszlo cites tends when informally understood to make it seem unlikely that anything at all will count as a system. If A is a part of B, one would think, any change in A would also be a change in B—else why call A a part of B? But this means that virtually anything, if it changes at all, will be bound to change *at least* as much as the sum of the changes of its parts. Since systems came about only on the condition that they changed "significantly" less than the sum of the changes of their parts, the Weiss-Laszlo point of view seems to ensure something of a short supply among the systems.[27]

Ordered wholes also evince deep powers of *adaptive self-organization*. This is the third of four marks of the systemic under the aspect of Laszlo:

We have shown that ordered wholes, i.e., systems with calculable fixed forces, tend to return to steady states following perturbations introduced in their surroundings. It is likewise possible to show that such systems *reorganize* their fixed forces and acquire new parameters in

[26] Paul Weiss, "The Living System: Determinism Stratified," in A. Koestler and J. R. Smythies, eds., *Beyond Reductionism* (New York: Macmillan, 1970). Weiss has one summation sign too many in the displayed formula.

[27] Of course, one often says that something or other is unchanged despite the obvious fact that its parts have changed. By this we generally mean that an object is unchanged in *essential* respects. Then, too, some devices are capable of minimizing certain variations in overall properties despite fluctuations in the magnitude of those properties locally; the human body's internal temperature generally stays within a degree or two of 98.6 even though the temperature at the surface may be higher. Neither characteristic has anything to do with adaptive self-stabilization.

their stationary states when subjected to the action of a physical constant in their environment.[28]

Evidence for adaptive self-organization is derived from W. Ross Ashby: "We start with the fact," Laszlo writes,

that natural systems in general go to ordered steady states. Now most of a natural system's states are relatively unstable. So in going from any state to one of the steady ones, the system is going from a larger number of states to a smaller. In this way it is performing a selection, in the purely objective sense that it rejects some states, by leaving them, and retains some other state by sticking to it. . . . The selection described by Ashby involves not merely the reestablishment of the parameters defining a previous steady state of the system after perturbation, but the progressive development of new steady states which are *more resistant* to the perturbation than the former ones.

Thus "every isolated determinate system obeying unchanging laws will develop 'organisms' that are adapted to their 'environments.' " From all this follows the "modified Ashby principle":

In any sufficiently isolated system-environment context, the system organizes itself as a function of maximal resistance to the forces which act on it in its environment.[29]

Other investigators, Laszlo adds evidentially, "have shown that, starting with an arbitrary form and degree of complexity . . . systems will complexify in response to inputs from the environment."[30] But ominously enough, the footnote alludes to Sewall Wright and S. A. Kaufman, biologists whose work in mathematical genetics and the computer simulation of evolutionary processes sheds roughly as much light on the notion of adaptive self-stabilization as the *I Ching*, which is to say, no light whatsoever.

But where the references are right, the argument is wrong, so the whole effect ultimately is unchanged, an achievement in perfect symmetry. Thus no investigator could have shown that all systems "complexify" in response to nudges from their environment, since many,

[28] Laszlo, *Introduction to Systems Philosophy*, p. 40.

[29] Ibid. W. Ross Ashby's *An Introduction to Cybernetics* (London: Chapman & Hall, 1956), a flabby and pretentious book, has had a sustained influence among systems analysts. It calls to mind, of course, Norbert Wiener's *Cybernetics or Control and Communication in the Animal and the Machine* (2nd ed., Cambridge, Mass.: The MIT Press, 1961), a book regularly hailed by freshman anthologists as one of the seminal works shaping and massaging the modern consciousness, and made remarkable by the number of integrals strung out on the page like so many linked sausages and pages and pages of soggy philosophy.

[30] Laszlo, *Introduction to Systems Philosophy*, p. 40.

such as the bacteria or the shark, do not. Some organisms develop as-tonishingly complex adaptations; but it is nonsense to affirm this as a *principle* of systems cybernetics. That living species do throw up adaptive variants is a brute *fact*, and one that requires explanation, but plenty of species never get adapted to anything at all, thrash around in the dark night of evolution, and then die out. This is behavior that has no digital analogue whatsoever, examples derived from Ashby notwithstanding.[31]

Nor is the argument strengthened by reformulation in terms drawn from information theory and thermodynamics:

Disorder in systems grows at a rate ds/dt. This is the dissipation function Ψ. Ψ may be positive, negative or zero. . . . If Ψ is negative, the system is in a state of progressive *organization*, that is, Ψ actually decreases its entropy or, what is the same thing, gathers information ($\Psi < 0 = d$ info$/dt > 0$).[32]

I pass over this without comment save to observe that from the last sentence one can infer that $0 \neq 0$, an evil omen, surely.

There is more. Having had wholeness, cybernetic stability, self-adaptation, and self-organization, we get the *inter-* and *intrasystemic hierarchies*, this in virtue of the fourth and final systemic property which guarantees, apparently, that

in systems which constitute ordered wholes, adaptively stabilizing themselves in their environment around existing steady states as well as evolving themselves to more adapted, and normally more negentropic (or informed) states, development will be in the direction of increasing hierarchical structuration.

Evidently this conclusion, whatever it may mean, follows hard on H. A.

---

[31] Thus Laszlo attributes to Ashby the following argument:

Suppose the stores of a computer are filled with the digits 0–9. Suppose its dynamic law is that the digits are continuously being multiplied in pairs and the right-hand digit of the product is going to replace the first digit taken. Since even × even gives even, odd × odd gives odd, and even × odd gives even, the system will "selectively evolve" toward the evens. But since among the evens, the zeros are uniquely resistant to change, the system will approach an all zero state as a function of the number of operations performed. (Ibid., p. 42)

But if the system's "dynamic law"—neither dynamic nor a law, so far as I can tell—were additive and not multiplicative, the system would "evolve" toward nothing whatsoever. It is hard to see, then, how examples of this sort support the astonishing conclusion that systems inevitably "complexify" in response to their environments. For some remarks on evolution generally, see my review of Ruse's *The Philosophy of Biology*, in *Philosophy of Science*, Vol. 41, No. 4 (December 1974).

[32] Laszlo, *Introduction to Systems Philosophy*, p. 44. The notation leaves something to be desired, of course.

Simon's hypothesis "according to which complex systems evolve from simple systems much more rapidly if there are stable intermediate forms than if there are not."[33]

Actually, when we turn to Simon's essay, "The Architecture of Complexity," what we find is not so much a hypothesis as a free-form example concerning fictional watchmakers, whose point appears to be simply that processes subject to periodic interruption proceed faster if they do not have to return to an original starting point than if they do—an observation on all fours with the discovery that relay racers are apt to win more races if they are not obliged to return to the starting post after each quarter of a mile. And just about as useful for an analysis of the theory of evolution since the analogy between partially completed watches and partially completed species is exceedingly dim.[34]

Whatever the moral behind the message from H. A. Simon, concepts of natural hierarchies are "entailments of the concept of self-stabilizing and self-organizing ordered wholes in common environments."[35] But the hierarchies, it turns out, defy rigorous explanation; instead, Laszlo writes, in his own limpid vernacular, that "the concept of 'hierarchy' ... will denote a 'level structure' or a 'set of superimposed modules' so constituted that the component modules of one level are modules belonging to some lower level." Suppose, Laszlo explains,

a given system a has component subsystems $c_1, c_2, \ldots, c_n$ in determinate association, the sum of which is expressed as R. System a is part of a hierarchy when its components are likewise systems and when it itself is a component of a more encompassing system

$$[a = (c_1, c_2, c_3, \ldots, c_n)^R] \subset b,$$

where a is a subsystem (component) of b, and all c's are comparable systems in relation to which a is a suprasystem.[36]

This is not quite a definition of what it is to be a hierarchy, since the definiendum seems to figure prominently in the definiens. Still, one sees what Laszlo has in mind and only the purist, I suppose, will observe that the formalism introduced above is strictly speaking meaningless. Systems have parts but they are also parts of other systems: various systems forming parts and wholes of each other are what Laszlo seems to understand by the natural hierarchies. There is no arguing this, what-

[33] Ibid., pp. 47–48.
[34] See H. A. Simon, *The Sciences of the Artificial* (Cambridge, Mass.: The MIT Press, 1969), pp. 90–95, for details. References to Simon are quite as common in the literature of GST as references to Ashby.
[35] Laszlo, *Introduction to Systems Philosophy*, p. 48.
[36] Ibid.

ever the construal of systemhood, but plainly set-theoretical inclusion does not suggest the relationship of part to whole: a can be part of b without being included in b. Uncalled for also is Laszlo's insistence that the resulting hierarchies are of the "Chinese box" variety, and are "theoretically infinite," especially since his own example seems finite.[37] It is this scruple that leads Laszlo unnecessarily to modify the natural hierarchies so that they are bounded above and below by basic and ultimate levels.

As it turns out, Laszlo brings to the natural hierarchies some skeptical concerns of his own. The "hierarchy of rigorously inclusive . . . natural systems," we learn, "is akin to Gerard's hierarchy of 'orgs' "; good news, surely, and evidence that if

a hierarchy of this kind could be empirically confirmed, a basic ideal of science would be realized: the many entities investigated by the diverse empirical sciences would be plotted on a map of hierarchical organization and the theories applicable to them could thereby be interrelated.

But the trouble is that "such confirmation encounters serious difficulties: the identification of different levels with empirically known entities is often problematic or unclear."[38]

I myself would have argued that it is the concept of a hierarchy itself that is problematic or unclear; perhaps this is a form of rhetorical excess, but in view of their alarmingly thin empirical content, things look tough for the hierarchies any way they are taken.

[37] Ibid.
[38] Ibid.

Isomorphisms:
Under the Aspect of Rapoport

Isomorphisms are the third of the three pledged usufructs of GST. The master's own discussion of these matters, with its notions of analogies, homologies, and explanations, remains more or less incomprehensible to me. However, the concept and the program to which von Bertalanffy alludes are well known. Isomorphism or structural identity is set-theoretical but not general: no single definition applies indifferently to *any* set-theoretical entity; rather, isomorphism holds between groups, or models, or rings, or the like. Typically, mathematicians and logicians will put a definition of isomorphism to use in the fashioning of *representation theorems*; these affirm the existence of a class of models such that every model of a given theory turns up isomorphic to members of this class. Cayley's theorem in group theory asserts, for example, that every group is isomorphic to a group of transformations.

Isomorphism is a concept useful not only within mathematics but as a notion figuring in the formal development of various sciences:

When a branch of empirical science is stated in exact form, that is, when the theory is axiomatized within a standard set-theoretical framework, the familiar question raised about models of the theory in pure mathematics may also be raised for models of the precisely formulated empirical theory. . . . Many of the discussions of reductionism in the philosophy of science may best be formulated as a series of problems using the notion of a representation theorem. For example, the thesis that biology may be reduced to physics would be in many people's minds appropriately established if one could show that for any model of a biological theory it was possible to construct an isomorphic model within physical theory.[39]

Certainly, important work has been done under this rubric. The reduction of thermodynamics to statistical mechanics is an acknowledged triumph of mathematical physics. But what remains in all this of GST? To discover that one model is isomorphic to another is only to diminish the stock of what was thought to be novel: isomorphism is an indicator of indifference. More to the point, discovering a significant representation theorem is hardly a task *external* to a given discipline. The reduction of thermodynamics to statistical mechanics was, after all, a triumph of mathematical physics; Cayley's theorem belongs specifically to group theory.

For all that, isomorphisms, epimorphisms, monomorphisms, mesomorphisms, endomorphisms, and the elusive *homomorphism*, continue

[39] Suppes, "Meaning and Uses of Models," p. 18.

to bulk large in GST.[40] Anatol Rapoport, for example, believes that GST "is not a 'theory' in the sense in which most scientific theories are theories," since its purpose is "to prepare definitions and hence classifications of systems that are likely to generate fruitful theories."[41] This does not bring Rapoport flatly into conflict with von Bertalanffy, who holds that GST is not a theory in the ordinary sense owing to its vastly greater generality, but their positions, described as antipodes, define a flat curve in space.

The usual notion of isomorphism or *structural identity* then leads to a "classification of all systems that can be represented by mathematical models." The "logical advantages" of such a classification "are at once apparent":

For instance, one mathematical system can be seen immediately as a generalization of another, that is, as including the latter as a special case. The induced classification of corresponding concrete systems immediately displays one class as including the other. If the systems are represented by isomorphic mathematical systems (or by the same model), all the theorems of the mathematical system are applicable to all consequences derived from the definition of the concrete system.[42]

The first sentence of this passage is false; the second a non sequitur; and the third without sense. But what is especially worrisome for GST taken under the aspect of Rapoport is not the general program but the confinement of concepts to particular cases. "Two mathematical systems are said to be *isomorphic* to each other," Professor Rapoport begins, "if a one-to-one correspondence can be established between the elements of one and those of the other and if all the relations defined on the elements of one hold also among the corresponding elements of the other."[43] As it happens, two mathematical systems may be isomorphic even if they share *no* relations; the definition to which Rapoport aspires—voiced for simple algebraic structures or models now —is this:

[40] See, for example, Klir, "The Polyphonic GST," p. 3.

[41] Anatol Rapoport, "The Uses of Mathematical Isomorphism in General Systems Theory," in Klir, *Trends in General Systems Theory*, p. 44. Rapoport cleaves to any number of alarming views. "A mathematical system," he writes some sentences before his definition of isomorphism, "is contentless. For instance, the set of positive integers closed under the operation of addition applies equally well to camels, to oranges, and to years." This, of course, is good evidence that we count with the integers, but hardly reason for supposing that arithmetic is devoid of content. If it were, the fact that we come up with the same results every time we add—whatever the objects being added—would be inexplicable.

[42] Ibid., pp. 46–47.

[43] Ibid., p. 46.

$M = \;<D, * > $ is isomorphic on D and D' to $M' = \;<D', \circ >$, where D and D' are sets and $*$ and $\circ$ are binary operations, if and only if there is a one-to-one mapping $F: D \to D'$ such that $F(x*y) = F(x) \circ F(y)$.[44]

But Rapoport writes as if the mathematical *theories* were isomorphic— "Isomorphism between two mathematical systems induces a conceptual isomorphism between the concrete systems that they represent."[45] — and thus in a master stroke collapses the entire discussion.

The reconstructed Rapoport, of course, means to talk of *categorical* theories, the most interesting being the theory of the positive integers, axiomatized according to the second-order Peano axioms, and the theory of the real numbers. Categorical theories are in sparse supply beyond the obvious examples, and in any case the definition of isomorphism can be framed only for structures of the same cardinality; given complex models of the sort required to model Newtonian mechanics, it is no easy task to decide whether pairs of models are isomorphic.

These are abstruse issues; I mention them in a prophylactic spirit. Thus consider as a case in point Rapoport's claim that the following two equations are isomorphic under the correspondence given at (1.8c):

$$m \frac{d^2x}{dt^2} + r \frac{dx}{dt} + cx = f(t), \tag{1.8a}$$

$$L \frac{d^2q}{dt^2} + R \frac{dq}{dt} + C^{-1}q = f(t), \tag{1.8b}$$

$$x \leftrightarrow q, \quad m \leftrightarrow L, \quad r \leftrightarrow R, \quad c \leftrightarrow C^{-1}. \tag{1.8c}$$

[44] See A. G. Howson, *A Handbook of Terms Used in Algebra and Analysis* (London: Cambridge University Press, 1972), p. 34, for a clear definition of isomorphism, epimorphism, monomorphism, and homomorphism.

[45] Rapoport, "Mathematical Isomorphism in GST," p. 47. Not to be outdone, Professor Klir writes that "the similarity in the forms of algebraic or differential equations is a kind of *mathematical isomorphism*. When this is generalized to include any relation, whether expressible by equations or not, then the concept of a *general system* acquires its proper meaning" ("Polyphonic GST," p. 2). If *equations* are being paired off, it makes no sense to talk of generalizations to *relations*—and vice versa. This is correctable nonsense depending on nothing other than the ability to confuse the use and mention of a term or set of terms. But directly after the sentence I have quoted, Klir goes on to say that a general system "is a contentless (mathematical) *representant* (*model*) *of a particular equivalence class*, obtained when an isomorphic relation (which is always an equivalence relation) is applied to certain traits of systems," and thereby achieves perfect incoherence.

The first equation, suitably interpreted, describes a damped harmonic oscillator; the second, an electrical circuit containing an inductance, a resistance, and a capacitance in series. This suggests *analogies* to Rapoport "between position and charge, between mass and inductance, between friction and electrical resistance and between elasticity and capacitance,"[46] but he wobbles in his appraisal of their intuitive plausibility once he gets past friction and resistance. The trouble is the oleaginous isomorphism. Structural similarity is a relation between models (or other algebraic structures)—but of *which* models can isomorphism be affirmed given the information at (1.8)? Surely not the set of *all* models that satisfy either (1.8a) or (1.8b) under their respective physical interpretations; *neither* equation determines its models up to isomorphism.

---

[46] Rapoport, "Mathematical Isomorphism in GST," p. 48. Professor Hempel considers a similar case, namely, a comparison of Poiseuille's law in fluid dynamics $[v = c \cdot (p_1 - p_2)]$ and Ohm's law for the flow of electricity in a metallic conductor $[I = R \cdot (V_1 - V_2)]$:

Thus the analogy in virtue of which the flow of a fluid here constitutes a model of the flow of a current may be characterized as follows: A certain set of laws governing the former phenomenon has the same syntactical structure as a corresponding set of laws for the latter phenomenon; or, more explicitly, the empirical terms . . . occurring in the first set of laws can be matched, one by one, with those of the second set in such a way that if in one of the laws of the first set each term is replaced by its counterpart, a law of the second set if obtained; and vice versa. Two sets of laws of this kind will be said to be *syntactically isomorphic*.

This, in turn, is used to make sense of the notion that a given system S_1 is an analogical model of another system S_2:

S_1 is an analogical model of S_2 with respect to the sets of laws L_1 and L_2 [if and only if] the laws in L_1 apply to S_1, and those in L_2 to S_2, and if L_1 and L_2 are syntactically isomorphic.

See Carl Hempel, *Aspects of Scientific Explanation* (New York: The Free Press, 1965), p. 436. This is somewhat clearer than Rapoport, but an obvious objection to the entire concept of syntactic isomorphism is simply that the same law or set of laws may be expressed by sentences that are *not* syntactically isomorphic. Thus consider two expressions for the familiar law of falling bodies:

$$\delta \int_0^T \left(\frac{1}{2} x^2 - gx \right) dt = 0 \tag{i}$$

$$\frac{d^2 x}{dt^2} = -g. \tag{ii}$$

Suppose we say, following Hempel, that (i) describes the system S_1 and (ii) describes S_2. Then, plainly, S_1 is not an analogy of S_2, a remarkable enough conclusion since S_1 and S_2 are in fact one and the same system, in any sense of system, however loose.

In fact, isomorphism is the wrong notion for what Rapoport is after: (1.8c) establishes nothing more than similarity in *form*, and this between the equations at (1.8a) and (1.8b) and not the models in which they may happen to be specified. Both are instances of the general second-order equation

$$\frac{d^2 x}{dt^2} = a_1(t)\frac{dx}{dt} + a_2(t)x + b(t). \tag{1.9}$$

It is interesting, perhaps, that many different phenomena can be described by equations of this form—with (1.9) multiply interpreted to talk about position and charge, mass and inductance, friction and resistance—but not remarkable: *any* differential equation can be expressed in the form

$$\frac{d^N x}{dt^N} = f\left(t,\ x,\ \frac{dx}{dt},\ \dots,\ \frac{d^{(N-1)}x}{dt^{(N-1)}}\right);$$

this certainly does not imply that all processes that change over time are structurally the same.

Form is in any case a slippery guide to content. The equations

$$\frac{dy}{dt} = y^2 \quad \text{and} \quad \frac{dy}{dt} = t^2 \tag{1.10}$$

match up item by item in the fashion of (1.8a) and (1.8b); only the second is *linear*, and linearity marks a stern segregation of mathematical properties.

Equations (1.8a) and (1.8b) in fact share more than form: they are both second-order linear differential equations *with constant coefficients*, instances alike of

$$\frac{d^2 x}{dt^2} = a_1\frac{dx}{dt} + a_2 x + b(t), \tag{1.11}$$

and it is *this* underlying sameness of mathematical structure that makes possible the switch from springs to circuits: second-order linear differential equations with constant coefficients have precisely the same solutions.[47] These equations make up a class of wide usefulness in the

---

[47] See G. Birkhoff and G. Rota, *Ordinary Differential Equations* (Waltham, Mass.: Blaisdell Publishing Co., 1969), pp. 83–92, for the details.

analysis of tuning forks, the pulsations of electrons in atoms, the interactions between hosts and predators in certain ecologies, as well as damped springs or the oscillations of charge flowing through a circuit.[48] I yield to none in my amazement that Nature has chosen (1.11) as an especially suitable system of mathematical relationships. *Soli Deo Gloria*. For all that, however, it remains difficult to see that the ubiquity of relations that might be expressed by (1.11) requires anything approaching a distinctive intellectual celebration.

[48] See R. Buck, "Logic of General System Behavior," in H. Feigl and M. Scriven, eds., *Foundations of Science and the Concepts of Psychology and Psychoanalysis* (Minneapolis: University of Minnesota Press, 1956).

Some Affable Disciplines

It is here, at this point, clearly
stated what, precisely, a system is.
There is no longer any question.
A system is a 6-tuple such that
and so forth. . . .

A. Wayne Wymore[49]

General systems theory is a discipline of unusual academic congeniality.
The chemical physicist takes pains to distinguish himself from the physical chemist, but the systems theorist looks affably on propinquitous
disciplines, arguing that when viewed rightly, general systems theory
includes cybernetics, mathematical systems theory, graph theory, set
theory, automata theory, recursive function theory, and information
theory.[50] Cybernetics, of course, is an artificial creation, like Esperanto,
made up of the theory of linear servomechanisms, information theory,
the theory of nerve networks—incongruent chunks, really, serving somehow to suggest to social scientists the heft and feel of a separate science.
The remaining sciences make up bits and pieces of a new intellectual
technology that has lumbered into prominence since the end of World
War II. Other subjects generally reckoned a part of this new intellectual
technology are probability theory, statistical decision theory, and all of
finite mathematics accepted as a single indigestible lump.

Systems researchers are limited by distinctive liabilities in moving
toward these affable disciplines. I mention four. In the first, a solid
subject is attached somehow to concerns that are manifestly preformal:
the general effect is power without purpose. In the second, the grand
aims of theory are couched in the logician's passionately scrupulous
language, but with no sense at all for its meaning and nuance. In the
third, serious if narrow subjects, such as information theory, are invoked
for systematic purposes they cannot serve. In the fourth, research results of painful modesty are introduced as solutions to problems of
great depth and complexity. I pass to appropriate illustrations.

[49] A. Wayne Wymore, "A Wattled Theory of Systems," in Klir, *Trends in GST*,
p. 276.
[50] Affable disciplines are on display in many current collections. There is, for example, Walter Buckley's *Modern Systems Research for the Behavioral Scientist*, a
collection of coughs in the night; George J. Klir's *Trends in GST*; Ervin Laszlo's
The Relevance of GST; Fred Emery's *Systems Thinking* (Baltimore: Penguin Books,
1969); William Gray and Nicholas D. Rizzo's *Unity Through Diversity: A Fest-
schrift in Honor of Ludwig von Bertalanffy* (New York: Gordon and Breach, in
press); M. D. Mesarović's *Views on General Systems Theory* (New York: John
Wiley & Sons, 1964); or Paul A. Weiss's *Hierarchically Organized Systems in Theory
and Practice* (New York: Hafner, 1971).

(1) An interesting affable discipline is recursive function theory. This is the domain scouted by Lars Löfgren in his paper "Relative Explanations of Systems." Remarkable in its provisionings, the paper sports two immense appendixes: there the theory of recursive functions and effective computability is expounded. What remains gives voice to Professor Löfgren's desire "to interpret the concept of general systems theory . . . as a common scientific language . . . a metalanguage in which we can discuss the effects of various logical bases for effective explanations of certain types of systems phenomena."[51]

Löfgren's chief thoughts are set out as hypotheses: the first is that "if an explanation E is effectively understandable (relatively effective), then it is understandable in terms of the rules . . . and axioms . . . which constitute an r-formal system S such that E is a p-explanation in S."

Now a formal theory, roughly, is one amenable to axiomatization; from the perspective of the recursive functions, formal theories are sometimes considered symbols spun off into finite lists according to well-defined rules of formation and derivation. A given finite sequence is called a proof if all transitions are rigorously rulelike. Formal systems resemble computer programs; computers, Turing machines. Turing machines are abstract devices suited to the computation of anything computable. Löfgren's p-explanations are proof sequences; his first hypothesis suggests that an explanation is effectively understandable if it occurs strung out in the context of a formal system. But I would argue that understandability is obviously a *preformal* concept (with the same degree of conceptual interest as the perfectly formed egg).[52] The botanist eager to learn of results in the theory of plate tectonics will not count an explanation ineffective because it has not been forwarded in a formal theory. On Löfgren's account the relatively effective explanations have been merged with those framed in a formal theory, but the ordinary content to the notion of a relatively effective explanation has disappeared, leaving behind only a chalky residue.[53]

---

[51] Lars Löfgren, "Relative Explanations of Systems," in Klir, *Trends in GST*, p. 343.
[52] See the first chapter of Hartley Rogers's *Theory of Recursive Functions and Effective Computability* (New York: McGraw-Hill, 1967) for an account of as much of recursive function theory as anyone would ever wish to know.
[53] Recursive function theory is an unusual but not a Rococo example of an affable discipline. Professor J. A. Goguen marks the contrast by calling on *category theory* as an instrument of analysis in his article "Mathematical Representation of Hierarchically Organized Systems," in E. O. Attinger, ed., *Global Systems Dynamics* (New York: John Wiley & Sons, 1970), pp. 112–129. A category is basically a collection taken together with certain operations on its members. The axioms that they satisfy make it easy to think of the objects belonging to a given category as themselves composed of highly structured entities: thus there is the category of all

(2) Professor Mihajlo Mesarović contributes his mite toward the analysis of systemhood in a paper entitled "Foundations for a General Theory of Systems." His aim is to achieve "a linguistic definition" of a
general system, and he begins with "the concept of a formal statement
in a language L," an inversion, I take it, of the logician's "statement in
a formal language L":

A *formal statement* F in a language L is defined as a sentence which is
formed according to the rules of grammar of the respective language
but the truth of which is not revealed by the statement itself. It is assumed that the formal statement has some unspecified constituents and,
consequently, the formal statement might be taken to be true for some
values of these constituents.[54]

This construction of the formal statements does not go a long way
toward distinguishing them from any of the others: save for the logical
trivialities, one cannot tell by inspection which of the grammatical
sentences of ordinary language are true and which are false. But Mesarović's second sentence also involves an obvious mishmash since the entire
point of formal languages is to avoid unspecified constituents of any
kind in favor of an array of symbols drawn from a list that is usually
fixed and finite. What Mesarović means, no doubt, is that a formal statement may contain *variables* variously satisfied and that a statement
might be true for some values of the variables and false for others.
This is sound enough for some statements, but false for others that
feature no variables whatsoever, but simply parameters, constants, and,
say, predicate symbols (Fa, for example, where a denotes a fixed integer
and F denotes the predicate "is a prime").

Assume, Mesarović continues,

that a set of formal statements K is given. If a subset M of these statements is taken to be true, it defines *a theory T over K*. Namely, theory
T conjectures that only the statements in subset M are true, the remaining statements being left unspecified.

Now this has an attractively rigorous ring, but the logician will see in the
quotation only a strong series of solecisms. Thus the phrase "a theory
T over K," which calls to mind the algebraic "vectors over the real
field" or the number-theoretic "primes over a division ring," is quite
without sense. A theory, on the usual logical view, comprises a con-

---

groups, the category of all vector spaces, and so on. Conspicuously absent from
the theory is anything like an interesting stock of theorems. Professor Goguen does
not disappoint: his is a paper filled with virtually contentless, inaccessible sym-
bolism.

[54] Mihajlo Mesarović, "Foundations for a General Theory of Systems," in Mesarović, *Views on GST*, p. 6.

sistent set of sentences, nothing more. Every consistent set of sentences is satisfied in some model, so for every theory we can, if pressed, speak of a model in which that theory is specified. Models are not, while theories are, strictly linguistic entities. A complete theory is an ordinary theory taken together with its logical consequences. The theory of a model M is just the smallest set of sentences satisfied in M and closed under the operation of logical consequence.

No better is Mesarović's definition of a general system as "a set of proper statements," since it suggests that the first n sentences of any formalized axiomatization of the propositional calculus constitute a general system.

Nor is the explicit definition of systemhood a great improvement. On Mesarović's view, a system is simply a *relation* on abstract sets:

$$S \subset X \left\{ v_i : i \in I \right\}, \qquad (1.12)$$

where X denotes the Cartesian product of the sets indexed by I. Any set is a subset of something, so (1.12) collapses the notion of a system into the notions of a sequence and a set.

(3) Often logical vulgarity—in the style of Mesarović, for example— and a nonsensical philosophy of science come together, exquisitely. In a piece entitled "Systems and Their Informational Measures," W. R. Ashby remarks that "the study of parts (in 'classic' science) must be supplemented by the study of wholes, [and] also that there *exists* a science of wholes, with its own laws, methods, logic, and mathematics."[55]

Modern systems science resists the dissective passion of the classicists. In addition to wading forthrightly into the nonlinear, systems theory demands an active confrontation with the grossness of complexity: "It not merely studies systems with high internal interaction but also confidently tackles systems in which it is the interactions themselves that are of interest."[56]

Appropriate tools for such systems, Ashby believes, are a mix of information theory and modern set theory. Take *relationships*:

The Bourbaki school has shown abundantly how every complex relationship may be regarded (rigorously) as corresponding to some *subset of a product* (Cartesian) *space*.

[55] W. R. Ashby, "Systems and Their Informational Measures," in Klir, *Trends in GST*, pp. 78, 79.
[56] Ibid., p. 80.

But this brings in information theory at a gallop since

as a *sub*set, the relationship represents a selection and thus joins precisely to information theory, which concerns itself largely with measuring the intensities of selections.[57]

The concepts that Ashby actually introduces are, however, remarkably
informal:

The basic idea of the method is that, with a system of n variables,
$X_1, X_2, \ldots, X_n$, the entropy of each one, $H(X_i)$, can be measured
directly on that variable without reference to the others. If we also
know the conjoint behavior (of the whole, as a vector), we can estimate the entropy of the system, $H(X_1, X_2, \ldots, X_n)$. Then the total
transmission between the n variables is defined by

$$T(X_1 : X_2 : \ldots : X_n) = H(X_1) + \ldots + H(X_n) - H(X_1, \ldots, X_n).$$

This total transmission, a measure of the departure of the whole from
internal statistical (probabilistic) independence, can then be analyzed
quantitatively in various ways so as to throw light on the various internal relations between the parts.[58]

As it happens, Ashby is interested not in the transmission capacity of
a given system, but rather in the system's *least safe capacity*—a quantity
that measures the transmission capacity of a system under the assumption that all internal probabilities are equal. Ashby correctly observes
that a system's least safe capacity "is seldom of interest in engineering
and infrequently mentioned in that field."

I bring the matter up only to illustrate a loop in Ashby's own thinking. It appears that the least safe capacity of a system, once full interaction between parts is countenanced, is apt to be gargantuan:

When an experimenter is trying either to study a system or to control
it, how much will the informational quantities be increased if the system is changed from one having no interaction between its parts to
one having full interaction between them?
    What *will* happen cannot be predicted . . . but the least safe capacity
can be found unambiguously. Suppose that the system has n variables
. . . and suppose also . . . that each variable can take k values. . . . Consider first the set of systems with *no* interactions between the variables.
In this set, the property holding for each variable . . . is *not* conditional
on the values of the other variables. Hence the subset, corresponding
to the relation embodied by the system, is some *rectangular* subset. . . .

[57] Ibid., p. 81. It is true, of course, that from a set-theoretical point of view any
relation may be thought of as a set of ordered pairs; this hardly calls, one would
think, for the Bourbaki. For good discussions of information theory see, for example, George Miller, "What Is Information Measurement?," or Anatol Rapoport,
"The Promises and Pitfalls of Information Theory," both contained in Buckley,
Modern Systems Research for the Behavioral Scientist.
[58] Ashby, "Systems and Their Informational Measures," p. 82.

The number of such rectangular subsets can easily be reckoned: each variable can provide 2^k subsets of *its* values, and n of such variables provide . . . 2^{kn}. For the selection . . . of one of them, the least safe capacity required is its logarithm to base 2, that is, kn bits.

When interaction between the parts is allowed without restriction, the system may embody a relation corresponding to *any* subset of the space. . . . As the whole space has k^n points, its subsets number $2^{(k^n)}$, and the selection, or identification, of one of them requires (as least safe capacity) k^n bits.[59]

This is opaque, yet the basic observation is simple. Given n distinct sets $S_1, S_2, \ldots, S_n$ of k members apiece, the available subsets for each S_i is fixed at 2^k; the sum of such subsets is $n(2^k)$. These subsets, in turn, determine a space of precisely $(2^k)^n$ possible relations. By way of contrast, the union $S_1 \cup S_2 \cup \ldots \cup S_n$ of n distinct sets of k members apiece determines a space of $2^{(k^n)}$ possible relations.

So what?

The point, it seems, is that systems with interacting elements have least safe capacities exponentially larger than systems with isolated elements: the least safe capacity required to fix a given relation in an isolated system is kn bits, while the capacity needed to fix a relation in the larger system is k^n bits:

Clearly we may say that allowing interaction causes an *enormous* increase in the magnitude of the least safe capacity. And we can see that, when full interaction is occurring, when the system is a "whole" in the fullest sense, no mere doubling or trebling of resources . . . is likely to be of any use.[60]

Now these are observations that have always prompted the ever-dissective classical scientist to ask why he should bother with systems taken as wholes in the first place, a full interactive account being such a palpable impossibility. It is against these unholistic conclusions that I expect Ashby to set his face: after all, information theory was introduced precisely to bring order out of the studied density of systems in their concrete wholeness. It is wonderful, then, and not a little surprising to read that

systems theory is essentially a demand that we treat systems as wholes, composed of related parts, between which interaction occurs to a major degree. No one supports this demand more willingly than I do, but the examples given above [of systems with staggeringly many possible relationships] show that, having won our battle for the admission of interaction, we must now learn moderation. . . . Systems theory, having broken away successfully from the extreme "classic" attempt to treat the whole as consisting of isolated parts, cannot go to the other ex-

[59] Ibid., p. 83.
[60] Ibid., p. 84.

treme, in which the interactions are total. . . . The future of systems theory, therefore, seems to lie in the study of systems that are sufficiently connected to be real systems yet by no means totally connected.[61]

(4) H. A. Simon is a systems theorist of slightly sturdier sophistication. *Models of Man*, his best-known work, has had a steady influence on social scientists; it is to this volume that they have repaired in making the case for the mathematical subtleties of systems theory.[62] A more recent text, *The Sciences of the Artificial*, records Professor Simon's Karl Taylor Compton lectures at MIT and devotes itself chiefly to such topics as artificial intelligence, the theory of design, and the heuristics of problem solving. By far the best essay is the last; entitled "The Architecture of Complexity," it was composed before Professor Simon delivered the Compton lectures and so appears tacked onto this collection as an editorial afterthought.

The theory of artificial intelligence represents a perfect partial instantiation of GST, embodying as it does some successful affable disciplines —recursive function theory, computer programming, automata theory— and involving an extended research program of dubious means, extravagant assumptions, and disappointing performance. The original premises favoring a theory of artificial intelligence were simple enough. Computers can execute any theoretically calculable task: insofar as human beings engaged in chess, checkers, or the routine construction of mathematical theories are doing anything that involves calculation, it should be possible to reinterpret their behavior sequentially and transcribe the results into a program for an actual or hypothetical computing device.[63] Physically realized, such a program would thus exhibit artificially intelligent behavior: machines that embodied it would act in a fashion that could not be distinguished from a purely human achievement.

This way of looking at cognitive acts makes possible the view that both human beings *and* computers are simply information-processing devices. In fact, much of the research in artificial intelligence is not directly tied to the computer, but deals more grandly with the creation of general information-processing strategies.

This is a crude but serviceable account of the fundamental idea, which

[61] Ibid., p. 85.
[62] S. Andreski has discussed *Models of Man* with happy ferocity in *The Social Sciences as Sorcery* (London: Andre Deutsch, 1972).
[63] For a clear statement of this thesis, see Allen Newell and H. A. Simon, *Computer Simulation of Human Thinking* (The Rand Corporation, P-2276, April 20, 1961); for a general exchange, see the section entitled "Man the Machine" in D. Berlinski, ed., *The Cutting Edge* (New York: Alfred Publishing Co., 1976).

dates roughly to the 1950s and in this country at least is bound up with such names as Allen Newell, J. C. Shaw, and H. A. Simon.[64]

Basic to the development of information-processing theories has been the annunciation of predictions of stunning optimism, largely in the absence of anything very much like experimental evidence. Thus Professors Newell, Shaw, and Simon claimed in 1957 that "we now have the elements of a theory of heuristic . . . problem solving," adding brightly that

we can use this theory both to understand human heuristic processes and to simulate such processes with digital computers. Intuition, insight and learning are no longer exclusive possessions of humans: any high speed computer can be programmed to exhibit them also.[65]

Professor Simon, at any rate, was so convinced by his own remarks that he saw in the near warp of time a computer that would play championship chess and derive high-class theorems.

That was in 1957. None of the predictions have come true; so uniform have the failures been, when measured against published projections— whether in machine translation, pattern recognition, adaptive control and programming, problem solving, or general heuristics—that some British researchers have taken to wondering whether the entire field might not in fact be empty, an analogue, perhaps, to napropathy or the principles of EST.[66]

A great deal of the work undertaken in artificial intelligence rests on a fundamental error. From the fact that programs can be written to simulate any human computation, it hardly follows that human beings employ such programs themselves, any more than human beings use the rules of transformational grammar to utter correct sentences, or a falling projectile computes an integral in order to arrive at the minimum distance it must travel. For almost all complex human activities, the evidence, however fragmentary and inconclusive, suggests strong *global* powers that do not reflect discrete, combinatorial, or sequential steps at all. No doubt *some* sequential programming exists for any of these

[64] For a survey of results in this field, see E. A. Feigenbaum and J. Feldman, *Computers and Thought* (New York: McGraw-Hill, 1963).

[65] Allen Newell, J. C. Shaw, and H. A. Simon, "Report on a General Problem Solving Program," in *Proceedings of the International Conference on Information Processing* (Paris: UNESCO, 1960), p. 257.

[66] See the account of the Lighthill Report in *Science*, Vol. 180 (June 1973), p. 1352. For an account of the field from an uncompromisingly negative view, see Hubert L. Dreyfus, *What Computers Can't Do* (New York: Harper & Row, 1972). See also F. J. Crosson, ed., *Human and Artificial Intelligence* (New York: Appleton-Century-Crofts, 1970) and K. R. C. Schan and K. M. Colby, eds., *Computer Models of Thought and Language* (San Francisco: W. H. Freeman, 1971).

activities, but global information-processing systems might be matched to programs of unfathomable complexity.

It is against this austere background that one must read Simon's *The Sciences of the Artificial*, especially since the plan of the second chapter is to convince the reader that

> *a man, viewed as a behaving system, is quite simple. The apparent complexity of his behavior over time is largely a reflection of the complexity of the environment in which he find himself.*[67]

This is a claim that anyone familiar with the actual as opposed to the projected progress of the social sciences will find remarkable.

As it happens, Simon means to restrict himself to man considered strictly as a cognitive device: his goal is to hit on those "simple principles that underlie human behavior"; he surveys some of the recent research in the theory of information processing and reports that

> the evidence is *overwhelming* that the system is basically serial in its operation: that it can process only a few symbols at a time and that the symbols being processed must be held in special, limited memory structures whose content can be changed rapidly.[68]

And this suggests

> the sorts of generalizations about human thinking that are emerging from the experimental evidence. They are simple . . . just as our hypothesis led us to expect. *Moreover, though the picture will continue to be enlarged and clarified, we should not expect it to become essentially more complex.*[69]

---

[67] Simon, *The Sciences of the Artificial*, p. 25.

[68] Ibid., pp. 52-53.

[69] Ibid., p. 53; emphasis added. The experimental evidence to which Simon is partial deals with the memorization of nonsense syllables, crude problem-solving strategies, and the like and involves capacities that are at once profoundly trivial and distinctly unlike the central cognitive abilities of human beings. Even so, he seriously overestimates the degree to which we have a solid sense of such subjects as the organization of human memory. In this regard, see Patrick Suppes, "On the Theory of Cognitive Processes," in Suppes, *Studies in the Methodology and Foundations of Science*, pp. 394-410, esp. p. 407. For a recent survey of results in this general area, see Steven W. Keele, *Attention and Human Performance* (Pacific Palisades, Calif.: Goodyear Publishing Co., 1973).

In dealing with grammars and the transformational grammarians, Simon does treat of topics that involve important intellectual properties, but he does so only by pressing his friendship upon his natural enemies. Against Chomsky he argues that the acquisition and execution of one's normal grammatical abilities reflect "characteristics of the human central nervous system which are common in all languages but also essential to other aspects of human thinking besides speech and language" (p. 48). Chomsky argues the contrary view in his recent *Reflections on Language* (New York: Pantheon Books, 1975), Chapter 1.

But these are generalizations only vaguely supported by the data, even the data restricted to the areas of information processing that Simon himself looks to as rocklike in favor of his thesis. Contrary evidence suggests complexity and a scientific grasp of phenomena that must in all reasonableness be called marginal.

No wonder, then, that when it comes to the extrapolation in favor of man viewed as an appealingly simple machine, more intricately wired perhaps than a computer but not appreciably different in kind, researchers with some real sense of the unspeakable complexity of the mind and the mysteries of its organization will look to Noam Chomsky:

There has been a natural but unfortunate tendency to "extrapolate" from the thimbleful of knowledge that has been attained in careful experimental work and rigorous data processing, to issues of much wider significance. . . . Experts have the responsibility of making clear the actual limits of their understanding and of the results that they have so far achieved. A careful analysis of these limits will demonstrate that in virtually every domain of the social and behavioral sciences the results achieved to date will not support such "extrapolation."[70]

The affable disciplines comprise a curious group. Information theory and automata theory possess independent content, however modest. Like game theory and decision theory, these are subjects with implications for the mathematical analysis of the social sciences. What made the appropriation of these disciplines as parts of GST easy to accept was the conviction that the social and political sciences, under the influence of strong mathematical formalisms, had commenced to gallop after the physical sciences shortly after World War II and were, for all anyone knew, galloping still. Seen from a strictly *internal* point of view, as specialists in each field might see them, the sciences that make up the new intellectual technology, despite many similarities in language, have distinctly different points of view. Cybernetics is a motley, automata theory a highly specialized branch of the theory of recursive functions; information theory deals with the very specific problems that arise when information is being passed along a channel of communication. The mathematical spirit is quite different as one passes from discipline to discipline: information theory is analytic; automata theory, algebraic. But from an *external* point of view, seen as scientifically minded spectators might see them, these disciplines are united by more than a common if promiscuous identification with GST. Under certain lights, they all seem to share a concept in the notion of a machine. Here Ashby strikes something of the right note:

[70]Noam Chomsky, *Language and Mind* (New York: Harcourt, Brace & World, 1968), p. XI.

Here we are obviously encroaching on what has been called General Systems Theory, but this last discipline always seemed to me to be uncertain whether it was dealing with physical systems, and therefore tied to whatever the real world provides, or with mathematical systems, in which the sole demand is that the work shall be free from internal contradictions. It is, I think, one of the substantial advances of the last decade that we have at least identified the *essentials* of the machine in general.[71]

Fixing the concept of a machine as central was a process that, in bulk, took place from 1947 to the present—at least in the social sciences. Psychologists had long seen opportunities for exploitation in the general notion, but not until the development of cybernetics, information theory, and the theory of automata did they feel secure in passing from exploitation to expropriation. The theory of automata made the mathematical notion of a machine explicit by linking it both to the theory of recursive functions and to the actual development of the digital computer. In information, information theorists hit on a powerful concept that was supposed to apply indifferently to a multitude of systems; and the omnivorous cybernetics, finally, seemed to provide a schema whereby machine behavior could be perceived as purposive and human behavior explained as machinelike.

The notion of a machine to which Ashby alludes, he also defines:

The machine with input or the finite automaton is today defined by a set of internal states I, a set of input or surrounding states, and a mapping, say, of the product set $I \times S$ into S. Here, in my opinion, we have the very essence of the machine; all known types of machines are to be found here; and all interesting deviations from the concept are to be found by the corresponding deviation from the definition.[72]

But his definition is not all that helpful. The restriction to finite and deterministic machines seems cramping even within the confines of automata theory. So long as what is wanted is basically a discrete device, the notion of a *Turing machine* provides a deeper and more general tool. A Turing machine consists of a finite set of states $q_0, q_1, q_2, \ldots, q_N$, a doubly infinite two-way tape scanned by a reading head and segregated into squares, and a finite series of symbols $S_0, S_1, S_2, \ldots, S_M$ that the machine is capable of printing, reading, and erasing. At any given time t_n, the machine is in some internal state q_i and the reading head is scanning a square on the tape. In moving from t_n to t_{n+1}, the machine can either halt operations entirely, erase the scanned symbol and change it for another, or shift the reading head from the

[71] W. R. Ashby, "Principles of the Self-Organizing System," in Buckley, *Modern Systems Research for the Behavioral Scientist*, p. 110.
[72] Ibid., p. 111.

given square to one adjacent, while changing its internal state. Instructions governing the change from t_n to t_{n+1} can be expressed as an ordered quadruple, $< q_i S_j R q_k >$, where q_i and S_j represent the initial state and scanned symbol, R a move on the tape, and q_k the new internal state. A Turing machine can then be *defined* as a finite, nonempty set of such quadruples.

A Turing machine can theoretically calculate anything calculable. This follows from an informal argument which invokes Church's thesis that calculability or computability collapses into recursiveness, and then shows the Turing machines to be capable of computing all and only the recursive functions. Nor do we need a clumsy plethora of such devices. The *universal* Turing machine can produce, as output, sequences embedded by any particular machine, thus achieving universal computing capacity.

But there is no real mathematical or moral reason to cut the machines off at a line that separates discrete from continuous devices. It is a cardinal axiom of the more modern systems theory that a system is simply the most general class of devices taking inputs to outputs by means of a set of states. When the transition functions are discrete, what results is an automaton; when they are continuous, one has something very much like a filter or continuous transducer.

The utility of these abstractions is hardly at issue. Automata theory generally, despite its somewhat uncertain and inelegant formulation, remains an exciting branch of recursive function theory, and the theory of continuous machines also flourishes.[73] But what of the larger hopes for the abstract concept of a machine touched on by Ashby and touted in GST? Roughly, they were three in number. First, the development of an abstract theory of machines was to provide a sophisticated way of construing complex human *cognitive* abilities. Machines would simulate human behavior, and the *theory* of machines would explain the analogous human capacities. Some theorists were emboldened to suggest that only the details were missing. Second, the development of cybernetics was to permit the perception of *purposive* behavior in machines, and thence to explain it *as* purposive so that human behavior formerly unreached by the machine analogy could finally be accommodated. Finally, the concept of information was to provide scholars with a tool of stunning generality, sufficient at the least for yoking together the various sciences. There were such obvious connections between infor-

[73] Recently, automata theory and recursive function theory have been manicured by some marvelously elegant mathematicians. As usual, such ministrations involve the recasting of known results in abstract and algebraic form. See, for example, S. Eilenberg and C. Elgot, *Recursiveness* (New York: Academic Press, 1970).

mation-theoretical and physical concepts of entropy, between information and probability.

John von Neumann expressed doubts about automata theory in 1951,
chiefy because it could not deal with continuous properties. His spirit
of skepticism has survived; it waxes rather than wanes. Machine translation is an acknowledged failure. Very serious and detailed work in linguistics has indicated human abilities that are unexplained in simple
automata theory or related disciplines. The entire automata-theoretical
enterprise of simulating human intelligence has swerved sharply from
its curve of ascending optimism. Even in relatively trivial areas such as
the construction of chess-playing programs, an impression of failure is
hard to avoid.

Something of the same thing has taken place in cybernetics and information theory. Both subjects are shaped about real theories, but
their applications have taken place in the paraplegic disciplines—sociology, psychology, political science, and management science—a sure
sign of debility. Cybernetics caused dispute from the beginning. Richard
Taylor argued that early formulations were philosophically objectionable, especially because they construed *purpose* as a behavioral concept.
One wonders now, some twenty years after the first popular accounts,
whether concepts borrowed from the uninteresting theory of servomechanisms have the force required to sustain cybernetics. Certainly,
the positive results have not been terribly impressive. Information theory is a much richer discipline than cybernetics, but much the same
limitations are now felt when it is pressed beyond the narrow problems
of communication channels for which it was designed.

All this is nothing new. Scholars have been expressing various sorrows
with these disciplines for at least a decade. But news travels with uneven
speed: the fine edge of skepticism that now characterizes linguistics
has not yet cut its way to various dark corners of the intellectual community. Worse, one sees a dismaying vulgarization of the theories as
they are stretched beyond limits of natural elasticity.

Evil Days for Systems Analysis:
Forrester's *Urban Dynamics*

But all this cannot possibly be appreciated in the abstract. The delight, as always, is in the details, and for these I turn to Jay W. Forrester's *Urban Dynamics*. Here is a fat book covering 250 pages and crammed with computer-theoretical arcana. Half the work is delivered to the reader in the form of a computer printout. Recondite charts dance across the pages; there are learned references to the DYNAMO compiler, pages and pages of densely printed input-output charts, and, finally, flow charts featuring intricately drawn arrows in numbers approaching the transcendental.

Urban Dynamics carries the ordinary systems-analytic hunger for the general to a point of baroque splendor, for in it Professor Forrester has assayed to explain the growth and decline not of any particular city, not even of a group of particular cities, but of urban areas *überhaupt*. Progress on this order has been formerly unobtainable, primarily because

the influences operating within a city are so subtly and intricately inter-connected that the human brain—whose response is conditioned by exposure to simple systems—finds it all but impossible to trace cause and effect.[74]

Professor Forrester, whose own brain has presumably smashed through the barrier of simple systems, has been sustained in his analysis by com-munion with the powers of systems theory:

The concepts of structure and dynamic behavior apply to all systems that change through time. Such dynamic systems include the processes of engineering systems, biology, social systems, psychology, ecology, and all those where positive- and negative-feedback processes manifest themselves in growth and regulatory action.[75]

Although Professor Forrester's model is of the twentieth order of complexity, the details can be summarized compactly. There is first of all the theory proper: a series of numbered equations that describe relationships between three economic classes, three classes of housing, and three kinds of business enterprises, all within a given area. The theory is segregated into what Professor Forrester calls *level* and *rate* equations. The former describe some magnitude associated with each sector at a particular time; the latter, associated rates of change. In

[74] William K. Stevens, "Computer Is Used as Guide for Expert Seeking Way Out of Labyrinth of Urban Problems," *New York Times*, October 31, 1969.

[75] Jay W. Forrester, *Urban Dynamics* (Cambridge, Mass.: The MIT Press, 1969), p. 1.

addition, *Urban Dynamics* contains a simulation run of the theory that fixes initial values for parameters and then traces predicted consequences over a period of 250 years.

Forrester's model is a hypothetical entity whose idealized features are presumed comparable, at least in major respects, with the features of virtually any urban area. Given an empty plot of land, various attractions and tugs first act to create a healthy urban area and then, over time and quite unmolested by outside influences, push the plot toward a state of stagnant equilibrium characterized by excessive underemployment, declining industry, and wretched housing. The depressing history of an urban area, of course, is completely determined by the theory—so the actual simulation run provides little in the way of surprises. The inevitable descent toward decrepitude is primarily, though not exclusively, a function of the ever-swelling poor (quaintly called "underemployed"), who come to the city first tentatively, and then in larger numbers, as public policies exacerbate the very conditions they were meant to alleviate. As the poor come, the middle class goes, antagonized by high per capita tax rates and frightened by their growing political impotence. As the middle class leaves, housing decays and industry wobbles. The completion of the process sees the city shuttered, stagnant, and slumly.

Naturally, Professor Forrester derives from his theory grave lessons for the conduct of government. Succoring the lower classes, however attractive over the short term, is an effort bound to end in disaster. Job programs, training programs, financial aid, tax subsidies, and low-cost housing construction all bulk large, weigh little, and achieve nothing. What is needed, instead, are techniques for grappling the middle and upper classes to the city's core, extruding the poor into the limitless environment beyond the city, and encouraging the growth of labor-intensive industries.

All this is advanced very tentatively at the beginning of the book, but with increasing confidence and assertiveness toward its end. Professor Forrester is undismayed by the absolute lack of evidence adduced in support of his theory: he dismisses the matter with untroubled dispatch. But despite its air of confident *Brie*, faith in both the accuracy and intelligibility of Forrester's theory, and the remarkable and astonishing claims that it entails, is morbidly affected by even a moderately close reading of the text.

Consider, for example, the first equations:

$$UA_{kl} = (U_k + L_k)\ (UAN)\ (AMMP_k),$$
$$UAN = 0.05,$$

$$(1.13)$$

where

UA = Underemployed arrivals (men/year)
U = Underemployed (men)
L = Labor (men)
k = Current year
kl = Period from k to l (1 year)
UAN = Underemployed arrivals normal (fraction/year)
$AMMP$ = Attractiveness-for-migration (of underemployed) multiplier
          perceived (dimensionless).

Equation (1.13) is the rate equation describing the arrival of underemployed into the area. In a given year the number of underemployed
males attracted to an urban area is assumed equal to a fixed percentage
of the total male population of the city, multiplied by a *perceived* attractiveness factor which would be unity under normal circumstances.
The *actual* attractiveness multiplier is given by

$$AMM_k = (UAMM_k)\ (UHM_k)\ (PEM_k)\ (UJM_k)\ (UHPM_k)\ (AMF),\quad (1.14)$$
$$AMF = 1,$$

where

$UAMM$ = Underemployed-arrivals-mobility multiplier
UHM = Underemployed/housing multiplier
PEM = Public-expenditure multiplier
UJM = Underemployed/job multiplier
$UHPM$ = Underemployed-housing-program multiplier
AMF = Attractiveness-for-migration factor.

All these factors are dimensionless.

Equation (1.14) states that the attractiveness of a given urban area
for a given class (in this case the underemployed) can be computed as a
product of six terms, each dealing with a different dimension of attractiveness and each receiving a value from a separate multiplier or function. The multiplier terms are related to relevant social parameters
through guessed-at curves.

These multipliers are something like utility functions defined to reflect not only preference but the *degree of preference* between alternatives. Multipliers are indicators of relative difference between the city
and its environments; the value of a multiplier, ranging from zero up-
wards, represents a cardinal weight assigned to a decisional factor, with
conditions of indifference represented by unity.

Equations (1.13) and (1.14), then, represent the main factors that account for the number of underemployed attracted to a given area. Together they constitute at least a partial explanation of what might be called *net propensity to migrate*. But even at first cut, much of this small theory is odd. The mathematics, such that it is, is purely ornamental. There is no more reason to assume that attractions are aggregated by the product function, as in (1.14), than to assume them aggregated by a function that extracts the cube root of the product multiplied by itself. And then, ought not net propensity to migrate be expressed probabilistically? Aggregate behavior is notoriously insusceptible to deterministic analysis. Surely a more sophisticated version would have it that variations in attractiveness affect the *probability* that the poor will choose to migrate. Additionally, and still *en passant*, one wants to know why UA is constructed as a function of the *size* of already existing working and underemployed classes. This feature is repeated throughout the equations that predict and explain propensities to migrate, and it seems equally arbitrary everywhere.

Forrester's equations account for no changes in arrival rates without the tacit assumption that men move toward urban areas they perceive as relatively attractive. So far, so good. But Forrester also assumes that agents only perceive as attractive those areas that *are* attractive, thereby leaving untreated the important case of false belief concerning relative attractiveness. More importantly, Equation (1.14) defines attractiveness itself along dimensions narrow enough to destroy the theory's psychological plausibility. Only factors of employment, housing, public expenditure, and occupational mobility get reflected. But urban migration in this century involves chiefly factors of land use—the decline of the agricultural way of life. Myriad other causes go uncited in the model. Consider such a simple parameter as *distance*. On Forrester's theory, two urban areas A and B should evince comparable rates of arrival if the right-hand sides of their respective model equations are equal. But what if the only available supply of underemployed agents is five miles from A and ten thousand from B?

You see the point, surely.

The model is quantitatively but not qualitatively sensitive to variations in the mix of attractiveness. This means that an urban area verging on chaos, but with a large population of workers and underemployed, might well be as attractive to the UAN class generally as one better managed but with a relatively smaller population of workers and underemployed. The total weighing construes identical products identically, a policy that wipes out differences in the *way* the products are determined. Nor do interactions between dimensions of attractiveness get reflected in the computation of aggregate attractiveness. But attractions are not generally independent: the total value of a plate of ham and eggs depends heavily on whether the ham and eggs or the plate is served first.

There are additional puzzles. Take, for example, the concept of normality. Multipliers translate relative attractiveness into numerical values that enter into the computation of total attractiveness at (1.14). The functions are arbitrarily pegged so that each multiplier has a value of unity when the argument reflects an equilibrium with the outside environment. Thus unity represents no net gain or loss in attractiveness. The curves described by the various multipliers are not linear, so there is ample room in each curve to reflect the fact that strictly equal increments in arguments do not necessarily produce equal increments in values. And this is as one would expect, for a 10 percent increase in public expenditures would have different attractive powers depending on whether it were added to a financial structure just comparable to, or vastly more wealthy than that of the environment. But, paradoxically, various *AMM* multipliers fixed at *different* equilibrium points all seem to be identical: their shape never changes. Suppose that there are two urban areas A and B interacting with two environments A' and B'. Imagine that equilibrium points at A are taken as Φ and at B as $\beta\Phi$, with $\beta\Phi > \Phi$. If the arguments of the multipliers ϵ and ϵ' are such that $\epsilon - \Phi = \epsilon' - \beta\Phi$, then, all other things being equal, the values of the multipliers and, hence, of the rates of migration turn out to be the same. More generally, any two cities with comparable degrees of relative attractiveness attract the same number of people. But this is silly. The behavior of people moving from misery to mere wretchedness is quite different from the behavior of people moving from luxury to absolute ravening opulence.

Naturally, in a theory covering almost 120 equations one cannot exhaust the possibilities for criticism in so short a space, but I would not want to ignore the rest utterly. There is, for example, on page 144, a record of the curious decision to fix departure rates of the poor as the *reciprocal* of inward migration rates. There is, on page 166, the hypothesis that managerial unwillingness to stay within a city is almost a linear function of rising tax rates—an assumption that makes the movement of managers toward the suburbs and higher per capita taxes hard to explain. On page 175, a strange connection is drawn between high tax rates and premium housing construction, one that leaves undiscussed the relationship between mortgage funds, labor costs, zoning restrictions, and depressed premium housing construction. Page 184 presents the astonishing assertion that a "history of successful housing construction, and the building industry organized to provide the construction, tends to *maintain* the construction rate" (my italics). On page 192, we read that "increasing managers in proportion to managerial jobs increases the likelihood of establishing new enterprise," a claim that predicts a peak of business dynamism during periods of mass unemployment. The very next page describes "the inclination to build new enterprise in terms of the availability of labor" and implies that a low labor/job ratio depresses, while an excessive ratio encourages, new construction. On page 218, we learn that as the underemployed population grows larger, it becomes politically more influential, consequently congratulating itself with higher tax expenditures while paying a disproportionately smaller share of the taxes.

The influence of GST is palpable here. Decision theory makes at least a partial appearance in Forrester's multipliers, for construed conveniently, they turn out to be utility functions defined for classes rather than individuals. Forrester, of course, must describe the fixing of the functions, as well as proposals for verifying their shapes.

But this is a quibble. The theory is really dominated by assumptions that belong quintessentially to GST. Virtually all important properties of an urban area, Forrester assumes, can be *explained* by describing processes and structures occurring within the urban area itself. Thus the environment collapses to an abstract point functioning solely as the source of men and relata for the relative inequalities that power the theory's multipliers; agents appear or disappear on the city's tape only when levels (or *internal states*) sink or rise. This thesis immediately puts one in mind of GST, especially in its automata-theoretical guises: cities are systems, systems are machines. More particularly, cities are *goal-directed* systems, so GST gets reflected in its cybernetic roles as well.

Not only are cities systems, they are amenable to study by *general principles* of systems good everywhere and for all systems. These principles are hinted at in *Urban Dynamics* and expounded more fully in a separate text entitled *Principles of Systems*. The theory gets plotted in Chapter 4, devoted exclusively to the *structure* of systems. However, when one attends closely to details, one finds little in the way of explication. The notion of *feedback* is never fully explained. Evidently, positive feedback is simply a barbarism denoting growth, while negative feedback has something to do with servomechanisms. But one cannot be sure. Terms like "decision" and "decision process" get dragged in without much explanation:

As used here the decision process is one that controls any systems action. It can be a clear explicit human decision. It can be a subconscious decision. It can be a governing process in biological development. It may be the valve and actuator in the chemical plant. It can be the natural consequences of the physical structure of the system. Whatever the nature of the decision process, it is always embedded in a feedback loop. The decision is based on the available information; the decision controls an action that influences the system level; the new information arises to modify the decision stream.[76]

Connoisseurs will want to read this paragraph backward as well as forward.

After siphoning off the merk we are left with something rather traditional. Forrester is chiefly interested in behavior that changes over time. His level equations are simply recordings of magnitudes associated with one or another physical quantity. The rate equations reflect changes in magnitude: the apparatus that is actually developed is nothing more than the traditional method of handling changes through time by means of differential equations. The principles of systems that were to hold universally turn out to involve nothing more than a clumsy application of the calculus.

All this might leave untouched the central thesis that cities are systems. And there is a point of abstraction at which this view can be rendered trivially true. If all laws needed to explain the city turn out recursive, then one can simply concoct a Turing machine to compute all and only the functions associated with those laws. But this is a mere flight of fancy, and there is no assurance that the internal states so uncovered will correspond in any way to states of the city itself.

[76] Jay W. Forrester, *Principles of Systems* (Cambridge, Mass.: Wright-Allen Press, 1969), p. 44.

2 Dynamical Systems

The full set of dynamic equations of a given physical system presented in one of the approximate forms, along with the corresponding boundary conditions and with the *algorithm for the numerical solution* of these equations—inevitably containing means from a finite-difference approximation of the continuous fields describing the system—form a physico-mathematical *model* of the system.

Andrei S. Monin[1]

Ordinary Differential Equations and Dynamical Systems

The theory of differential equations—one of the three great differential theories—comprises the mathematical technique for the analysis of change. Given typically is a system of differential equations,

$$\frac{dx_i}{dt} = f_i(t, x_1, x_2, \ldots, x_n), \qquad i = 1, 2, \ldots, n, \tag{2.1}$$

where the integration of f_i expresses the continuous but varying relationship between changes in time and changes in x_i. A differential equation has not one but a *family* of solutions; if (2.1) meets certain straightforward conditions, the existence and uniqueness of the system's solutions are guaranteed for fixed initial values of the equations' parameters and functions.

I have urged the distinction between theories and their models; here, too, there is some danger of seeing systemhood in the differential equations themselves—a danger exacerbated by the common but misleading habit of calling a string of differential equations a "system." *Dynamical systems* correspond to differential equations in roughly the same way that models correspond to theories.[2] The family of solutions to an ordinary differential equation could plausibly be taken as the chief example of a dynamical system. If the solutions depend continuously on variations in the initial data, the dynamical system is *well-posed* or *well-set* in the sense of Hadamard; well-posed problems receive their most extensive analysis in the development of classical physics, with the most important physical example of a dynamical system afforded by celestial mechanics.

A certain notational uniformity can be enforced by equating dynamical systems to *differential models*. A system's *state* is the smallest collection of numbers that must be specified in order to predict uniquely the system's behavior; at (2.1), each state is a point in an n-dimensional

---

[1] Andrei S. Monin, *Weather Forecasting as a Problem in Physics* (Cambridge, Mass.: The MIT Press, 1972), p. 117.

[2] For an account of the theory of dynamical systems, see J. K. Hale and J. P. La Salle, eds., *Differential Equations and Dynamical Systems* (New York: Academic Press, 1967), especially the articles by M. M. Peixoto and S. Smale. Also, G. S. Jones, ed., *Seminar on Differential Equations and Dynamical Systems* (New York: Springer-Verlag, 1968). For the theory of ordinary differential equations, a standard account is E. A. Coddington and N. Levinson, *Theory of Ordinary Differential Equations* (New York: McGraw-Hill, 1955).

Euclidean space. The set of all such points comprises a *phase* or *state space*. The various x_i can be considered components of a vector; f_i defines a *vector field*. A dynamical system, then, is a differential model consisting of a phase space together with its associated vector field.[3]

The study of differential equations—and the associated models in which they are satisfied—splits into quantitative and qualitative branches. There is the business of simply finding the system's solutions or, failing that, approximating those solutions numerically—this is the province of the quantitative theory. But most interesting differential equations cannot be solved in terms of standard functions, or even in terms of those functions taken together with various integrals (an example is $dx/dt = x^2 + t$); and only a computer can fully carry out the various schemes for numerical integration.

Quantitative work has its special dangers in that spurious moral grandeur is generally attached to any formulation computed to a large number of decimal places. Faced with an authoritative avowal that pig-iron production is increasing at an annual rate of 13.295674 percent, only the incorrigible skeptic will raise the possibility that pig-iron production may not be increasing at all.[4]

The qualitative theory is still under the influence of Poincaré, who created the subject in a memoir of 1881.[5] Fundamental is the geometrical point of view; since for most systems of differential equations analytic solutions are impossible, what becomes important instead is the *phase portrait*, a study of the geometrical character of trajectories representing the range of solution graphs. That differential solutions have an essentially geometrical meaning makes it possible to assess their analytic properties indirectly, by reckoning directly with their geometry.[6]

---

[3] Alternatively, a dynamical system might be defined as a one-parameter group of transformations $f(p, t)$ ($-\infty < t < +\infty$) of a space $\Omega(p \in \Omega)$ into itself satisfying the following conditions:
1. $f(p, 0) = p$;
2. $f(p, t)$ is continuous in p and t;
3. $f[f(p, t_1), t_2] = f(p, t_1 + t_2)$.
This would be a dynamical system defined in the fashion of Nemytskii and Stepanov. For details, see V. V. Nemytskii and V. V. Stepanov, *Qualitative Theory of Differential Equations* (Princeton, N.J.: Princeton University Press, 1956).
[4] René Thom has written interestingly on this matter in "Modern Mathematics: An Education and Philosophical Error," *American Scholar* (November-December 1971).
[5] Jules Henri Poincaré, "Sur les courbes définies par une équation différentielle," *Oeuvres*, Vol. 1 (Paris: Gauthier-Villars, 1928).
[6] See Solomon Lefschetz, *Differential Equations: Geometric Theory* (New York:

In the simplest case, the system of first-order differential equations collapses into a single equation,

$$\frac{dx}{dt} = f(t, x). \tag{2.2}$$

The ideal, of course, is to describe an explicit solution $x = u(t)$ governing changes in x as a function of changes in t. Failing that, the analyst looks to (2.2) for the characteristic geometric structure that it induces on the TX plane. To each point a short line segment of fixed slope $f(t, x)$ is assigned. A filled plane of this sort is called a *lineal* or *direction field*. Imagine now the plane filled with curves tangent at each of their points to the lines whose slopes are determined by the lineal field. Such curves themselves fill out the plane, and each embodies a differential solution since for any function $u(t) = x$ defined by one of the curves,

$$\frac{du(t)}{dt} \equiv f(t, u(t)) \tag{2.3}$$

for every t in the interval of definition of u.

Nothing much changes in the geometrical interpretation of (2.1): the lineal field of (2.2) passes to a *vector field*, but the induction of geometrical structure on an underlying space proceeds in just the same fashion. The system of equations

$$\frac{dx}{dt} = y,$$

$$\tag{2.4}$$

$$\frac{dy}{dt} = -x,$$

thus assigns to every point in the XY plane a vector $\{y, -x\}$; solutions arise geometrically as *trajectories*—the graph of points moving through the field whose velocity at (x, y), say, is simply $\{y, -x\}$.

John Wiley & Sons, 1963), for the best available introduction to the qualitative theory.

More generally, if (2.1) unfurled is

$$\frac{dx_1}{dt} = f_1(t, x_1, x_2, \ldots, x_n),$$

$$\frac{dx_2}{dt} = f_2(t, x_1, x_2, \ldots, x_n),$$

$$\vdots \qquad\qquad\qquad\qquad\qquad (2.5)$$

$$\frac{dx_n}{dt} = f_n(t, x_1, x_2, \ldots, x_n),$$

a solution appears as a curve in the n-dimensional Euclidean space $\mathbf{R}^n$:

$$x(t) = \{x_1(t), x_2(t), \ldots, x_n(t)\}, \qquad\qquad (2.6)$$

such that

$$\frac{dx}{dt} = \left\{ \frac{dx_1}{dt}, \frac{dx_2}{dt}, \ldots, \frac{dx_n}{dt} \right\}. \qquad\qquad (2.7)$$

The natural geometrical interpretation depicts the right-hand side of (2.7) as a vector tangent to the curve at t.

Nor need the underlying space be essentially Euclidean. In mechanics, Euclidean spaces involve unattractive restrictions; Poincaré favored the *differential manifold* as a more general space for the analysis of purely Hamilton vector fields.[7] The global point of view afforded by the qualitative theory enabled Poincaré to resolve decisively celebrated issues in celestial mechanics that had remained for a century within the realm of quantitative theory. The stability of the solar system is the most important case in point. The arguments of a long series of brilliant nineteenth-century figures traded on series-expansion techniques. But Poincaré showed that an important group of such series expansions diverge, and Bruns proved that methods other than series expansion would not resolve the n-body problem quantitatively; and so the issue stood until Poincaré, and later Birkhoff and Moser, using the topological methods

[7] For a superb discussion, see Ralph Abraham, *Foundations of Mechanics* (New York: W. A. Benjamin, 1967).

of the qualitative theory, were able to prove the existence of quasi-periodic solutions to the three-body problem. Such solutions, of course, remain quite beyond the reach of *any* computational techniques since they extend over the entire range of the phase portrait; nor has any analytical description of their fundamental properties been given.

Social Systems:
Under the Aspect of Forrester
and Meadows

Carthage was destroyed by the
Romans. This is called *system
destruction* or *dissolution*.

Morton A. Kaplan[8]

Classical physics owes much to the theory of ordinary differential equations, and systems of such equations figure conspicuously in mathematical economics too. But in political science and sociology there has been resistance to a recasting of theoretical dynamical principles in differential form, no doubt because neither discipline has anything very much like a body of theoretical principles to begin with. There is Richardson's work, of course, and recently some social scientists have turned to such subjects as linear control theory, optimal control theory, and the theory of differential games.[9] For the connoisseur of differential methods, however, linear control theory is disappointingly vulgar; in optimal control theory and the various theories of differential games, differential elements are subordinate since such sciences are really decision-theoretical in their overall cast. Pure differential theories have come again to prominence in ambitious works such as *The Limits to Growth* and *World Dynamics*. These are volumes of studied defects, much appreciated by their critics. But Forrester and Meadows do see human agents much as the physicist sees particles in a field of force—as brute and unanalyzable elements. So their work is valuable as an example of the unencumbered differential method. My own concern is less with either book than with the genre: like *World Dynamics*, the theories that make it up are global, highly aggregated, desperately nonlinear. Taken one with another, they form a collection of mathematical models of the world, and it is interesting to ask for a preliminary listing of representative points in the *natural critical space* in which they may be evaluated.

The Limits to Growth and *World Dynamics* are ambitious and sustained efforts to see in human and social systems the elements of a dynamical system amenable to description and analysis by means of differential equations.[10] They are gloomy and fashionable documents.

[8] Morton A. Kaplan, "Systems Theory," in James C. Charlesworth, ed., *Contemporary Political Analysis* (New York: The Free Press, 1967), p. 152.
[9] For an interesting account of differential models in the social sciences, see J. Gillespie and D. Zinnes, "Progressions in Mathematical Models of International Conflict," *Synthese*, Vol. 31, No. 2 (1975).
[10] Donella H. Meadows, Dennis L. Meadows, Jørgen Randers, and William W. Behrens III, *The Limits to Growth* (New York: Universe Books, 1972). This volume is a popular account of the authors' *Dynamics of Growth in a Finite World* (Cambridge, Mass.: Wright-Allen Press), scheduled for publication in mid-1973 but still

The Limits to Growth stands forward and just slightly to the left of Jay Forrester's *World Dynamics:* behind them both are the lessons of *Urban Dynamics* (A Study in Slums), *Industrial Dynamics*, and *Principles of Systems*, the eminences of applied dynamics.[11]

Things are bad and getting worse; by the end of the century a point of crisis will have arrived with all the inevitability of Death. This is the mostly Malthusian message of *The Limits to Growth* and *World Dynamics*. Professors Forrester and Meadows are voices made dolorous by the awesome powers of the exponential function:

Within one lifetime, dormant forces within the world system can exert themselves and take control. Falling food supply, rising pollution, and decreasing space per person are on the verge of combining to generate pressures great enough to reduce birth rate and increase death rate. When ultimate limits are approached, negative forces in the system gather strength until they stop the growth processes that had previously been in control. In one brief moment of time the world finds that the apparent law of exponential growth fails as the complete description of nature. Other fundamental laws of nature and the social system have been lying in wait until their time has come. Forces within the world system must and will rise far enough to suppress the power of growth.[12]

Nor do they hope for *technological* succor: the computer is a harsh taskmaster. Those splendid initiatives to which optimists habitually appeal—dynamic rice, colorful condoms, mulched plankton—prove languidly incapable on simulation of affecting the coming catastrophes to anything more than a marginal extent. The Ecological Evil One has so arranged the affairs of this world that a society that has successfully evaded disaster on account of failing natural resources is sure to encounter it as the result of intolerable overcrowding, gross pollution, or blighted crops.

This is not a position that Professor Forrester has reached as a mere

---

unavailable at the time of writing of this chapter. My comments are quite general, however, and so do not depend crucially on a close examination of the details of the theory as they are set out in *Dynamics of Growth in a Finite World*.

[11] Jay W. Forrester, *World Dynamics* (Cambridge, Mass.: Wright-Allen Press, 1971); *Urban Dynamics* (Cambridge, Mass.: The MIT Press, 1969); *Industrial Dynamics* (Cambridge, Mass.: The MIT Press, 1969); *Principles of Systems* (Cambridge, Mass.: Wright-Allen Press, 1968).

There are two notable collections of criticism that deal with *The Limits to Growth*. These are H. S. D. Cole, C. Freeman, M. Jahoda, and K. L. R. Pavitt, eds., *Models of Doom* (New York: Universe Books, 1973) and the report on *The Limits to Growth* issued by the World Bank. Both *Science* and *Nature* have run innumerable pieces on Forrester and Meadows; the reports in *Science* exhibit that magazine's usual mixture of solemn pomposity and peevish ignorance. See, for example, M. Shubik's piece in *Science*, Vol. 174 (1971), p. 1014.

[12] Forrester, *World Dynamics*, p. 5.

servant of fashion (*Hacke ordinaire*). His is no hastily and ill thought out document put together to satisfy a morbid popular taste for gloomy prognostications; it is a *theoretical* position, an apotheosis of Method. That *The Limits to Growth* and *World Dynamics* reflect dogmas grown momentarily popular is an accident of Science; only the skeptic will take seriously reports that the leading ideas for *World Dynamics* were composed in a matter of hours, as Dr. Forrester flew across the Atlantic after a meeting with sleek industrialists of The Club of Rome.[13]

His research, Dr. Forrester avows, has driven him to the information-feedback systems: "Everything we do as an individual, as an industry, or as a society is done in the context of an information-feedback system"—an object of uncommon conceptual congeniality that exists "whenever the environment leads to a decision that results in action which affects the environment and thereby influences future decisions."[14] Not surprisingly, "systems of information-feedback control" turn out to be "fundamental to all life and human endeavor."[15]

The feedback loop is what chiefly characterizes information-feedback systems; open systems, by way of contrast, involve nothing more than unconstrained linear lunges:

The feedback loop is a closed path connecting in sequence a decision that controls action, the level of the system, and information about the level of the system, the latter returning to the decision point.[16]

While the feedback loop is the basic element from which systems are created, its elements decompose further into *levels* and *rates*:

Within the loops of a system, the principles of systems tell us that two kinds of variables will be found—levels and rates. The levels are accumulations (integrations) within the system. The rates are flows that cause the levels to change.
A level accumulates the net quantity that results from the flow rates that add to or subtract from the level. The systems levels fully describe the state or condition of the system at any point in time. One's bank balance is a systems level. . . . Levels exist in all subsystems . . . [and] are caused to change only by the related rates of flow.
A rate of flow is controlled only by one or more systems levels and

---

[13] In any case, "To reject this model because of its shortcomings without offering concrete and tangible alternatives would be equivalent to asking that time be stopped" (Forrester, *World Dynamics*, p. ix). Citing the shortcomings of the model in just this fashion is indulging in "vague criticism about their lack of perfection," a temptation to which I find myself irresistibly inclined.

[14] Forrester, *Industrial Dynamics*, pp. 14, 15.

[15] Ibid., p. 15.

[16] Forrester, *Principles of Systems*, p. 1-7.

not by other rates. *All systems that change through time can be represented by using only levels and rates.* The two kinds of variables are necessary but at the same time sufficient for *representing any system.* [17]

The world model itself is expressed as a numbered string of finite-difference equations, the notation largely in the cumbersome DYNAMO style, a Stalinist mass of tightly bunched capital letters.[18] After concessions to the computer, what remain are five first-order, linear, autonomous equations in five monumentally aggregated variables. A second set of equations fixes the parameters of the first; the whole business is designed for numerical solution and simulation.

Although Professor Forrester believes, in the fashion of the unheralded pioneer, that his results are counterintuitive and intellectually inaccessible,[19] even the groping novice, looking on his bleakly plunging curves, can tell that something unattractive is in the works, and only the incorrigible optimist would declare himself astonished by the conclusion that industrial civilization as it is presently constituted is scheduled to expire with all of the inevitability and none of the grandeur that physicists associate with the final heat-death of the universe-as-a-whole.

World Dynamics spans the globe with five integral equations covering five states: population, capital investment, natural resources, fraction of capital devoted to agriculture, and pollution. The overall behavior of these variables is available in a graphical record "showing the mode in which industrialization and population are suppressed by falling natural resources." But the skeptic will want to know why these curves have just the shape that they do and no other, and so the issue is forced willy-nilly back to the symbol-laden theory again.

For at least part of the readout, *World Dynamics* has a straightforward theoretical justification: population, capital investment, and pollution all grow exponentially in the years between 1900 and 2000 because they are *inherently* exponential; in fact, "population, capital invest-

[17] Ibid., p. 1–9.

[18] The details of the DYNAMO Compiler are spelled out in Alexander L. Pugh's *DYNAMO II User's Manual* (Cambridge, Mass.: The MIT Press, 1973). For an account of the various simulation languages, see Howard S. Krasnow's "Simulation Languages," in T. H. Naylor, ed., *The Design of Computer Simulation Experiments* (Durham, N.C.: Duke University Press, 1969).

[19] In fact, he comes to this conclusion for purely theoretical reasons, holding that "with a high degree of confidence we can say that the intuitive solutions to the problems of complex social systems will be wrong most of the time. Here lies"— he goes on to add—"much of the explanation for the problems of faltering companies, disappointments in developing nations, foreign-exchange crises, and troubles of urban areas" (*Urban Dynamics*, p. 110).

ment, pollution, food consumption, and standard of living have been growing exponentially throughout recorded history."[20] So the basic axiom of the entire system lends itself to apothegmatic expression: *Between an irresistible impulse* (population growth) *and an immovable object* (fixed limits to natural resources), *something has to give.* There is no arguing with this in the abstract, but there is room for doubt over particular cases. (Thus Professors Brooks and Andrews have argued grandly that for all *practical* purposes the stock of mineral resources is infinite, a fact of only marginal comfort: one cannot eat zinc.)[21]

What general theory augments this axiom is given by a series of relationships variously modified by the system's multipliers. I trace the steps for that part of the system-as-a-whole governing the growth of population.[22]

There are two mutually interacting states to this system: population itself and the stock of natural resources. Natural resources are fixed and finite; only the rate at which they are consumed varies. The birth and death rates are both affected by the rate of resource consumption, and this interaction is expressed through a series of nonlinear multipliers. In turn, the birth and death rates affect the rate at which resources are being consumed, since the resource-usage rate is itself pegged to the level of population.

The disappearance of natural resources is determined by the *natural-resources-usage rate*,[23] which in Forrester's theory equals the population multiplied by some "normal" constant augmented by a special multiplier reflecting the increase in consumption that attends a higher standard of living. Thus, whatever the normal rates of resource consumption, the rate-as-a-whole is accelerated by increases in a population's size and wealth.

Given this, the present stock of natural resources can be computed by a simple integration. The fraction of the initial stock that remains is dubbed NRFR; it is this fraction that gives the crucial natural-resource-extraction multiplier, NREM, that traces the extent to which the efficiency of capital investment is reduced by a shrinking supply of natural resources. Figure 2.1 is read from right to left, as in Hebrew. The point (1,1) indicates a state of perfect adequacy of natural resources; points to the left mark the stages of depletion, with a parallel

[20] Ibid., p. 2.
[21] David B. Brooks and P. W. Andrews, "Mineral Resources, Economic Growth, and World Population," *Science*, Vol. 185 (July 5, 1974), pp. 13–19.
[22] See *World Dynamics*, p. 29.
[23] Ibid., p. 39.

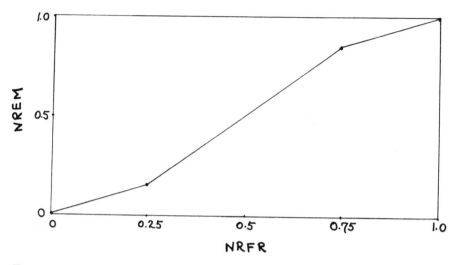

Figure 2.1
Natural-resource-extraction multiplier vs. natural-resource fraction remaining. From Jay W. Forrester, *World Dynamics* (Cambridge, Mass.: Wright-Allen Press, 1971), p. 37.

decline in capital efficiency (illustrating, no doubt, the maxim that *the less there is, the harder it is to get*).

The natural-resource-extraction multiplier is the chief determinant of the effective-capital-investment ratio, ECIR, a kind of alphanumeric wedge standing between NREM and the material standard of living, MSL. To obtain the ECIR, consider first the *effective capital unit*, obtained by discounting *all* capital by a factor that measures the portion of capital devoted to resource extraction. This calls to mind the NREM; in fact, the ECIR is determined by multiplying the available capital by the NREM. Clearly the values of the ECIR and the NREM will be paired: *The harder it is to obtain the goods, the less of the total capital stock each person gets.*

The material standard of living describes the extent to which effective capital investment per person is greater or less than the 1970 value, and variations in the material standard of living, finally, are translated directly into multipliers affecting birth *and* death rates (Figures 2.2 and 2.3).

Together with the natural-resource-extraction multiplier, these relationships are the pivots of the dynamical system.[24] They have a matched but uneven effect, with a rise in the standard of living leading to a decline in both birth and death rates but a decline giving rise to a modest

[24] Ibid., pp. 35–36.

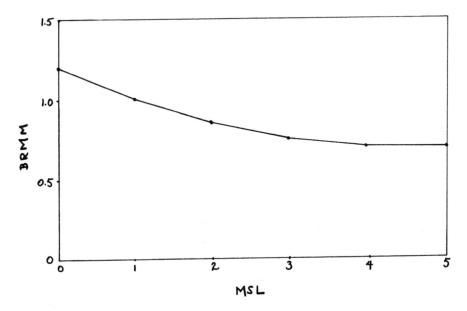

Figure 2.2
Birth-rate-from-material multiplier vs. material standard of living. From Forrester, *World Dynamics*, p. 35.

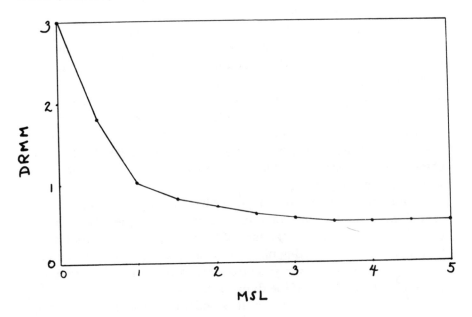

Figure 2.3
Death-rate-from-material multiplier vs. material standard of living. From Forrester, *World Dynamics*, p. 40.

increase in the birth rate and a *threefold increase* in the death rate.

Population, finally, is described by an expression of the form

$$P(t_k) = P(t_j) + \int_{t_j}^{t_k} [BR(t) - DR(t)] \, dt, \tag{2.8}$$

where P represents population, BR and DR the birth and death rates, and t_j and t_k the initial and final times for the computation. In other words, the present population is obtained by adding to the population at some previous time the souls who have come into the world and subtracting those who have departed.[25]

In the end, it is a decline in the world's store of natural resources that breaks the population boom and the accompanying expansive rate of industrial growth. This is achieved through the effect that a decline in natural resources has on capital investment and thereby on the world's standard of living, which is directly connected to the death rate.

Structure and Function

For the man of fastidious mathematical sense, an analysis extruding computational detail, couched in a boring boxy language, is an exercise in poor taste, like passing around samples of one's sputum.

B. L. Hendricks[26]

Professor Hendricks, I dare say, has overstated the case. But not by much. One uses the computer only from a purging sense of mathematical necessity. Some models are simply intractable: they have no analytical solutions. In such cases, the computer paces the theory through a series of steps—the simulation—with dynamical properties read directly from the numerical entries wherever possible. This is not easy. Numerical methods generally leave the *global* character of the solutions that they compute obscure, and it therefore becomes difficult to appraise the intuitive plausibility of proposed relationships.[27] What

---

[25] Forrester, *World Dynamics*, p. 33.

[26] B. L. Hendricks, *Black Particle Physics* (Emmantown, N.J.: The Emmantown Community College Press, 1972), p. 415.

[27] Occasionally, this makes numerical computation highly vulnerable, as in the case of systems of ordinary differential equations of the form $dx/dt = Ax$ (A a constant matrix) where the system has eigenvalues with both positive and negative real parts. In the system

$$\frac{dx_1}{dt} = x_2, \qquad \frac{dx_2}{dt} = 10x_1 + 9x_2,$$

one has to go on is simply a stretch of a curve whose outermost lineaments are not known. Planetary motion, for example, is governed by simple laws whose dynamical characteristics are not perfectly clear: although astronomers have no trouble predicting the course of the planets, the solar system may compose an unstable whole, set to fly apart on perturbation. Stability is quintessentially a qualitative concept of the sort that may not be obvious from a simulation run, no matter how elaborate.

In econometrics, theories have expanded at an alarming rate. Lawrence Klein has written of econometric models of 1000 equations or more; none actually exist, but many of the well-known models, such as the Brookings model or the Klein-Goldberger model, are complex enough to make simulation a practical necessity.[28] These theoretical instruments generate a matched pair of problems: to understand, on the one hand, a mass of economic data by means of the model, and to understand, on the other, the qualitative nature of the model by means of an appropriate simulation of the theory. In both cases, clearly, much is left unexplained: it might be a neat irony if the model turns out to be a less perspicuous object than the economy it was originally invoked to explain.[29]

---

for example, small numerical errors in computation, framed as the solution is being tracked toward zero, will yield a solution tending toward infinity. See Otto Plaat, *Ordinary Differential Equations* (San Francisco: Holden-Day, 1971), p. 155.

[28] See Lawrence R. Klein, "Forecasting and Policy Evaluation Using Large Scale Econometric Models: The State of the Art," in Michael D. Intriligator, ed., *Frontiers of Quantitative Economics* (Amsterdam: North Holland Publishing Company, 1971). For a recent discussion of some of the issues pertaining to the Brookings model, see Karl Brunner, ed., *Problems and Issues in Current Econometric Practice* (Columbus, Ohio: College of Administrative Science, 1973), especially the ripe exchange between Klein and Fromm and Basamann.

To some extent, techniques of numerical computation and simulation have made possible certain experiments on such large-scale objects as an economy or a corporation. See, for example, T. H. Naylor, ed., *Computer Simulation Experiments with Models of Economic Systems* (New York: John Wiley & Sons, 1971), especially the important article by E. Philip Howrey and H. H. Kelejian, "Simulation Versus Analytical Solutions," which suggests that analytical techniques—spectral analysis, for example—may ultimately prove as suitable for the examination of large macroeconomic systems as numerical techniques. A good discussion of spectral analysis can be found in T. H. Naylor, K. Wertz, and T. H. Wonnacott, "Spectral Analysis of Data Generated by Simulation Experiments with Econometric Models," *Econometrica*, Vol. 37 (April 1969), pp. 333-352.

[29] See, for example, Irma Adelman and Frank L. Adelman, "The Dynamical Properties of the Klein-Goldberger Model," in Arnold Zellner, ed., *Readings in Economic Statistics and Econometrics* (Boston: Little, Brown, 1968), for a sense of what is involved in determining the dynamical (or qualitative) properties of a model in

The equations that make up the theory of *World Dynamics* have none
of this complexity, despite an exposition suggesting inexpungible obscurities. In fact, the underlying theory is plain to the point of triteness.
Its chief assumptions, Meadows writes, are that
- population and capital by their very nature tend to grow exponentially;
- there are limits to the growth of any physical quantity on a finite
Earth.[30]

These are thoughts that lend themselves nicely to expression in the
austere notation of the theory of ordinary differential equations. Thus
the basic structure to the theory set out in *World Dynamics* is simple
enough: the system is pegged to a series of *differential equations*: it is a
system in virtue of being a *dynamical* system.

The state equations of *World Dynamics*, which serve as the chief instruments of systemhood, are integrations:

$$x(t) = \xi + \int_T^t f(s, x(s))\, ds. \tag{2.9}$$

But integral equations are equivalent to differential equations with
fixed initial conditions,

$$\frac{dx}{dt} = f(t, x), \qquad x(T) = \xi, \tag{2.10}$$

so there is no particular need to stick with (2.9), although integral equations do lend themselves more readily to simulation (simply because
integration is a smoothing operation, while differentiation is not).

There are, in all, five states or levels to the system of *World Dynamics*;
seven associated rates deal with births, deaths, natural resources, capital
investment, capital discard, pollution generation, and pollution absorption. Many of the equations depict halves of the same process and
should be combined. The *dynamical* character of the system is spread
uniformly through the equations; all processes that grow, do so in much

---

just 25 simultaneous equations. No really interesting theoretical object, of course,
is completely accessible. The field equations of general relativity still are not perfectly understood, some fifty years after they were announced.

[30] D. L. Meadows, "Typographical Errors and Technical Solutions," *Nature*, Vol.
247 (January 11, 1974). An additional assumption that Meadows mentions has to
do with the *lagged* nature of responses; since *World Dynamics* does not feature a
lagged structure, I do not mention it here.

the same way: by a fixed constant under *normal* conditions; by a constant boosted by various *multipliers* under conditions reflecting interaction between states. The multipliers are functions; in *World Dynamics* what explicit information there is about their characteristics is provided by a series of graphs—Figure 2.2, for example, where the birth-rate-from-material multiplier is plotted against the material standard of living and indirectly against natural resources. In fact, the multipliers are really functions of time and behave rather like arbitrary parameters, a fact which is not made clear in the text. In Figure 2.2, neither axis specifies the time, but both BRMM and MSL are fixed at unity under 1970 conditions. Thus for all *practical* purposes the relationship between the two is specified *only* for an arbitrary interval centered on 1970. To get a sense of the relationship between the birth rate and the material standard of living in, say, 1870, *another* point must arbitrarily be fixed as normal, and the *same* curves rerouted to that point. Thus a functional relationship between the two multipliers is generated only when one considers the entire class of curves taken at various points of normality or at various initial conditions. As it turns out, however, this resuscitative way of looking at the multipliers leads promptly to absurdity.[31] Structural relationships between multipliers are susceptible to variations over time; such relationships are not autonomous, and their representation must therefore include a temporal parameter.

Consider, for example, what is involved in the assumption that the relationship between the *natural-resource-extraction multiplier* and the *natural-resource fraction remaining* is independent of the time at which the stock of resources is being mined. Suppose a given population that is not pressing on the limits to growth were to increase dramatically. On the theory of *World Dynamics*, assuming now a set of functional relationships transposed to new conditions of normality, the rate of resource consumption would increase, the fraction of resources remaining would decline, and the material-standard-of-living multiplier would affect both birth and death rates. For certain reference points, this will be obvious nonsense: the *effective demand* for natural resources might be so low as to make an increase in population negligible.

Thus, in the larger sense of things, one must consider the multipliers to be only partially specified by the series of guessed-at graphs, with values beyond a given interval undetermined.

This is not to say that the internal organization of the equations that remain is now entirely clear. Forrester makes much of nonlinearity, and this is a feature one therefore expects in his treatment of the multipliers

---

[31] See D. Meadows, "Response to Sussex," in Cole et al., *Models of Doom.*

representing the factors that affect the birth rate, for example. Now the system of first-order differential equations in (2.1) is said to be *linear* if it can be expressed in the form

$$\frac{dx_i}{dt} = \sum_{j=1}^{n} a_{ij}(t)x_j + b_i(t), \qquad i = 1, 2, \ldots, n, \tag{2.11}$$

where the $a_{ij}(t)$ and $b_i(t)$ are continuous functions of t over some interval $t_1 < t < t_2$. If the $b_i(t)$ are all identically zero, the system is said to be *homogeneous*.

If we look closely at Forrester's theory, taken under conditions of "normality" in which all his multipliers are set to one, we find that the five state equations can be expressed in the classic form

$$\frac{dx_i}{dt} = a_i x_i, \qquad i = 1, 2, 3, 4, 5, \tag{2.12}$$

the a_i being constants. In the case of the population equation, for example, the constant is $a_{BR} - a_{DR}$ where the birth rate is $a_{BR}P$ and the death rate is $a_{DR}P$, P being the total population. This system of independent equations is obviously linear and admits of the analytical solutions

$$x_i = k_i\, e^{a_i t}, \qquad i = 1, 2, 3, 4, 5, \tag{2.13}$$

where the k_i are constants fixed by the initial conditions.

The system (2.12) is the crude core of the world model. Under the action of the system's multipliers, the theory swells internally to form a more complicated string of equations:

$$\frac{dx_i}{dt} = \left(\prod_{j=1}^{J} M_{ij}\right) x_i, \qquad i = 1, 2, 3, 4, 5. \tag{2.14}$$

Written thus, the system is recognizable in spirit, however bizarre in form. The choice of *multipliers*, one might note, is a fairly arbitrary way of including interactions among the equations. It is probably also worth stressing, if only because Forrester denies it, that (2.14) is still quite linear.

The system's appearance of dense internal complexity, which results from the collocation of multipliers in each of the five separate equations, is to a certain extent deliberately contrived to make plausible the overall methodological hypothesis that the world's economic and political relationships are of such rebarbative complexity that the intellect unassisted by the computer stands before the equations that describe them in a state of rapt befuddlement. The multipliers are, of course, functions whose arguments are functions in turn; these are the state variables covering population, natural resources, etc. The state variables are functions of *time*, so that changes in the values of the multipliers are ultimately pegged to changes in time, and this prompts an obvious simplifying measure. In order to decouple the system of equations at (2.14), make explicit the time-bound character of the multipliers at one fell swoop by defining for each state variable a separate temporal parameter. Thus if we let

$$A_i(t) = \prod_{j=1}^{J} M_{ij}, \qquad i = 1, 2, 3, 4, 5, \tag{2.15}$$

(2.14) admits of transmogrification to

$$\frac{dx_i}{dt} = A_i(t) x_i, \qquad i = 1, 2, 3, 4, 5, \tag{2.16}$$

a system of five independent, uncoupled equations. This is rather a radical restructuring of the theory's form. But if the simplification of (2.14) obliterates the system's inner structure, which is expressed entirely in the play of the various multipliers, it also reveals—since (2.14) and (2.16) are by definition equivalent—its fundamentally fatuous and excessively simple dynamical character. To the charge that the conversion of (2.14) to (2.16) represents spurious economy, since the values of the $A_i(t)$ cannot be determined until for any given t the theory's original multipliers have been fixed by simulation, I am unsympathetic. Values for the system's true multipliers are, in the context of the theory, given *administratively* through the guessed-at graphs and not derived from any set of compelling theoretical assumptions. So long as coefficients to the system's chief dynamical equations are determined in a fashion most charitably termed arbitrary, it makes good sense, and in any case satisfies some minimal urges toward mathematical simplicity, to capture *all* of the theory's gratuitous assumptions in a single param-

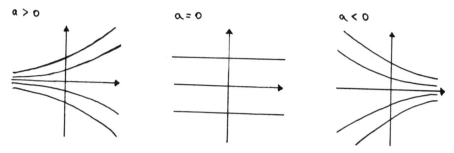

Figure 2.4
Behavior of the family of exponentials ke^{ax} for various values of a.

eter or functional coefficient, rather than having them spread awkwardly
over a mass of separate multipliers.

The whole of *World Dynamics* thus breaks into three parts. The first
two correspond to the system taken at either its normal or its augmented states; these are the systems at (2.12) and (2.16). The third
is purely administrative and consists in the determination, by means
of graphs or simple function tables, of the values of the system's multipliers.

The "normal" system, which in many respects corresponds to what
the engineer would call a free system (i.e., one in which forcing terms
are suppressed), comes about when in (2.16) $A_i(t) = a_i$ for all t in a
fixed interval. For a system so crude the qualitative structure to the set
of solutions is quite obvious. The exponentials of (2.13) form families
that are all monotonically increasing, monotonically decreasing, or at
equilibrium, depending only on the sign of a_i (see Figure 2.4).

The enriched core is hardly more mysterious. Indeed, the palpability
of solutions to the system—this in the qualitative sense—makes the
distinction between the theory's assumptions and its cardinal conclusions very hard to grasp.[32]

The system (2.16) can be solved formally by an elementary separation
of variables:

$$x_i = k_i e^{\int A_i(t)dt}, \quad i = 1, 2, 3, 4, 5. \tag{2.17}$$

This, to be sure, still leaves an expression that cannot generally be analytically investigated. However, the assumptions of the theory are strong

[32] Robert M. Solow makes the same (quite obvious) point in his "Notes on 'Doomsday Models,' " *Proceedings of the National Academy of Sciences*, Vol. 69, No. 12
(December 1972), pp. 3832-3833.

enough to suggest its general qualitative character for any particular form of the $A_i(t)$ for which the system might be investigated.

Consider, in particular, the state equation for population,

$$\frac{dx_P}{dt} = A_P(t)\, x_P. \tag{2.18}$$

We assume an initial value x_{P_0} at a time t_0, and we would like to find out as much as we can about the behavior of x_P over some interval I. Now the assumptions of the theory have it that the population is growing rapidly, that it will reach a limit, and that, having reached this limit, it will proceed rather inexorably to decline. This is not mathematically precise by any means, but the governing postulates do impose mathematical conditions on the equation's coefficients that for all practical purposes make the solutions quite trivial. Thus $A_P(t)$ *must* in the nature of things be bounded above and below by some A_P^{max} and A_P^{min}. Presumably $A_P(t)$ is also continuous and, within the limits formed by its upper and lower bounds, monotonic as well. There are thus no constant solutions to the population equation; those that do exist are all generally described in a qualitative way by remarking that as $A_P(t)$ approaches A_P^{max}, population size is, of course, increasing; at $A_P(t) = A_P^{max}$, growth is proceeding at its fastest *rate*; as $A_P(t)$ approaches zero, $x_P(t)$ approaches a maximum; and as $A_P(t)$ approaches A_P^{min}, $x_P(t)$ will approach a minimum value of some sort. So solutions show a series of ascending and descending curves: Increase—Approach a limit —Decrease. Or vice versa.

The skeptic with anything like an unforced and aristocratic contempt for the computer will naturally react to demonstrations of the system's dynamical transparency by asking, with just a touch of asperity, why the system's dynamical equations require an explicit solution by simulation, especially when what data exist inspire something less than perfect confidence in any of the very particular quantitative estimates required for the operation of a simulation run. Somewhat surprisingly, it is Professor Meadows himself who rises to just this point; in response to criticism that the actual values figuring in the various simulations are almost wholly nonsensical, Meadows argues boldly against the niggling parochialisms of the quantitative approach: The data are incomplete, he allows, but what information there is

[is] sufficient to generate valid basic behavior modes for the world system. This is true because the model's feedback loop structure is a much more important determinant of overall structure than the exact

numbers used to quantify the feedback loops. Even rather large changes in input data do not generally alter the *mode* of behavior. . . . Numerical changes may well affect the *period* of an oscillation or the *rate* of growth or the *time* of a collapse, but they will not affect the fact that the basic mode is oscillation or growth or collapse.[33]

Autonomy, underdetermination, overcompatibility

The vulnerability of *World Dynamics* to queries about time is already a matter of celebration. In examinations of the system's functional relationships it became apparent that their assured air of temporal omnivorousness was a mere sham: the functional relationships are plausibly specified only for the run of the simulation itself—that is, for about 100 years. In the theory of ordinary differential equations, those that are *autonomous* do not depend explicitly on time; the distinction is evident in the pair

$$\frac{dx}{dt} = f(x), \quad \frac{dy}{dt} = g(x, t). \tag{2.19}$$

Plainly the first will remain invariant with respect to transformations $t' = t + C$ of the time axis; the second, since it contains an explicit temporal parameter, will not. Autonomous equations are analytically more tractable than all the rest; this is the reason, I am sure, for their prominent position in both *World Dynamics* and *The Limits to Growth*. In autonomous systems, the structure to a set of relationships does not vary, although the relationships between variables may, of course, change over time. The great laws of physics are autonomous: the introduction of explicit temporal parameters into systems of equations is considered an embarrassment. In econometric models, however, much the reverse is true, and for good reason: structural relationships between economic agents *do* decay, especially since the relationships one has are apt to reflect nothing so magisterial as the laws of nature. This requires the introduction of dynamical equations that are not autonomous—precisely what one does not get in *World Dynamics* or *The Limits to Growth*.

The observation that a theory may frequently be underdetermined by the data available to judge it suggests another line of criticism. Indeterminacy of theory has been a philosophical staple over the years; Professor Quine has labored tirelessly on its behalf. Much the same theme has cropped up in econometrics as discussions of the *identification*

---

[33] Meadows et al., *Limits to Growth*, p. 121.

problem. Theories may be underdetermined by themselves as well as the data: this is a rare occurrence, something like a self-inflicted dueling scar, but evidence for it is available in *World Dynamics* and *The Limits to Growth*. A theory underdetermines itself when its chief functional relationships are mathematically unspecified. Thus, were one simply to *sketch* the relationship between x and y expressed by $y = f(x)$ for a fixed sequence of points within a narrow interval, while arguing grandly for an interpretation of $f(x)$ that gobbled up the asymptotes, one would have a textbook case of theoretical underdetermination, especially if the curves at hand represented nothing by way of a close fit to data culled from time series. Underdetermination is a liability, of course, because it leaves the global character of the theory indefinite.

What would render the multipliers explicit? Surely it will be insufficient to observe simply that they are continuous or differentiable. One possible line that for mathematical reasons I find attractive is that solutions to (2.16) are all essentially *sinusoidal*—functions of the form $Ae^{\alpha t} \cos(\omega t + \epsilon)$, where α is a damping coefficient. Seen this way, the Forrester equations fit into a well-known class of differential equations; and although nothing Forrester suggests justifies this construal, keeping damped oscillations in mind gives to the theory a kind of inner mathematical structure that fits in nicely with the empirical projections that Forrester and Meadows make. Thus a function that swings about a point of equilibrium with ever-decreasing amplitudes will, in the case of population growth, bring a given population past its natural limits on the upward side and then below these same limits on the downward side. This corresponds to Meadows's "overshoot and collapse mode," with the whole enterprise repeating itself on a diminished scale the next time around, the diminution corresponding to the fact that a population once whipped past the limits to growth will have far fewer members with which to rebound on the second oscillation. And so on, of course, until the oscillations die out and the population stays pretty much at equilibrium.

The fact that there are no overall mathematical constraints on the multipliers bespeaks a global indeterminism; locally, there is *overcompatibility*, with alternatives to the various graphs sustaining their qualitative spirit but specifying frankly incompatible solutions to the equations of the system. Consider the relationship between the death-rate-from-material multiplier and the material standard of living depicted in Figure 2.3. As the MSL drops toward zero, the DRMM rises sharply toward three; in fact, this stretch of the curve is virtually linear. Suppose, however, that the slope were changed slightly so that as the material standard of living dropped toward zero, the DRMM rose toward two, or perhaps approached three asymptotically along a flattened and somewhat shallow curve. These are changes within the qualitative spirit of the original curve. They reflect the overall gross assumption that a decline in the standard of living will raise the death rate, yet they have a significant effect on the curve that ultimately represents population growth: rather than rising to a limit and then changing to decline abruptly, the world population on these assumptions would rise slowly to a limit and then decline slowly. The appropriate curve would look rather like a flattened T square stood at oblique angles to the axis.[34]

Functional-differential equations

In an object pure as a system of differential equations, changes of state are registered immediately. This, of course, is an unnatural idealization; econometricians realize that effects *lag* behind their causes, and the study of such phenomena is by now a well-established cottage industry. To cite a single example, a widely studied class of functional-differential equations that take some account of lagging has the form

$$\frac{dx(t)}{dt} = f(t, x(t), x(t - r)), \tag{2.20}$$

where $x(t - r)$ reflects something very much like a prior state or *memory*

---

[34] To observations of this character Professor Forrester habitually replies that despite the variations, *some* form of catastrophe will overtake the growth of population. In *World Dynamics*, this is true: if a falling stock of natural resources does not bring population down, pollution or a crisis in the production of food will. But this is to misread the objection: what is at issue here is not the ubiquity of gloomy mechanisms, each calculated to induce widespread misery, but rather the provisionings that uniquely specify *one* such mechanism as being demanded by the theory.

Needless to say, objections on this order have nothing to do with estimations concerning the *truth* of such demographic projections as Forrester and Meadows make. In many respects the situation is even worse than Forrester and Meadows suggest. Using data taken from a United Nations demographic study, Forrester, Mora, and Amiot have argued that world population is growing at a rate faster than the exponential rate which is the limiting logistic rate! For their discussion, see "Doomsday: Friday, 13 November, A.D. 2026," *Science*, Vol. 132 (1960), pp. 1291-1295.

of the system. If $r = 0$, (2.20) goes over into an ordinary differential equation.[35]

In *Urban Dynamics*, delays are introduced by a distinction drawn between real and perceived states of the system; the two may differ, and changes in crucial state variables may therefore not crop up until such differences are noticed. There is nothing comparable in *World Dynamics*, and in any case such devices simply add an additional functional relationship to the system without lagging any of the variables in the fashion of (2.20).

Styles in simulation

Systems that have been simulated are vulnerable to two major sources of error: the first is the process of simulation itself, which like any numerical computation has particular liabilities; the second, the penumbra of uncertainty that surrounds the estimation of structural parameters. Schemes for numerical integration exist for both differential equations and finite-difference equations. In essence, and sticking now to differential equations, the computer will approximate a set of convergent solutions. In the very simplest of cases, the differential equation

$$\frac{dx}{dt} = f(x, t) \tag{2.21}$$

is evaluated by means of a function table that yields solutions at a sequence of points $t_0 < t_1 < \ldots < t_N$. Cauchy polygons are the simplest way to construct appropriate function tables: a partition or mesh is created on the interval of integration and then, using suitable Riemann sums, the values of the function table, given a particular initial value of the differential equation itself, are computed by the recursive formula

$$f_0 = c, \qquad f_k = f_{k-1} + f(x_{k-1}, t_{k-1})(t_k - t_{k-1}). \tag{2.22}$$

The approximate solution is then defined by means of a linear interpolation formula. What results, of course, will be the Cauchy polygon, a curve consisting of segments each of which is a straight line. Nothing much changes when difference instead of differential equations come into play: the underlying model begins with a suitable integral equation, which is then converted to appropriate difference form. The meth-

---

[35] See, for example, Kenneth L. Cooke's "Functional-Differential Equations: Some Models and Perturbation Problems," in Hale and La Salle, *Differential Equations and Dynamical Systems*, pp. 167–185.

od of Cauchy or Euler has an exact analogue here, and it is precisely this sort of numerical estimation that grounds both *World Dynamics* and *The Limits to Growth*.

These computations are subject to errors of *interpolation* and errors of *round-off, discretization*, or *truncation*. My guess is that round-off errors will not prove terribly important in estimating the accuracy of *World Dynamics*.[36] Interpolation errors may have some significance. But the sheer clumsiness of the simulation scheme is an irritation since *World Dynamics* is committed to the very simplest procedures in numerical computation. What is worse, in comparison to ordinary simulation systems the schemes of *World Dynamics* are decidedly liable to increases in interpolation errors. Here the point of comparison should be any of the improved Euler systems—the Runge-Kutta method, for example, or the system of Milne. It is an easy matter to prove these procedures considerably more accurate than those of *World Dynamics*.

Innocents of rigor
Both Forrester and Meadows are innocents of rigor (*naïfs statistiques*): missing from their work is some sense that a substantial body of statistical technique must mediate between an original theory and its applications.

Yet how much space, the skeptic will ask, *is* there between a theory and the data that it proposes to explain? On the conventional view, the answer is not much. "All swans are white," the zoologist observes, and it takes no more than a black swan to dispose of the matter. This might be considered the null case: zero distance between the assumptions of a theory and the data that either fit or fail to fit the view that the theory projects. But in the social sciences the available data are vast, the assumptions indirect and problematical, the connection between hypothesis and evidence difficult. Thus the advent of the metrical sciences—econometrics for economics; polimetrics for political science; biometrics for biology; even cliometrics for history—whose task is to interpose themselves between the broad-ranging and frequently untestable assumptions of a given theory and the mass of data the theory is meant to confront. Each of the metrical disciplines carries a charge of theoretical assumptions, chiefly about the relationship of theory to data; and so a confirmed theory within economics, for ex-

[36] But see H. S. D. Cole's "The Structure of the World Models," in Cole et al., *Models of Doom*, p. 30. For an account of techniques of numerical integration and the like, see Birkhoff and Rota, *Ordinary Differential Equations*, Chapters 7 and 8. For numerical methods generally, see R. W. Hamming, *Numerical Methods for Scientists and Engineers* (New York: McGraw-Hill, 1973).

ample, that has been reinterpreted econometrically reflects not one but a number of theoretical views. This is a point insufficiently stressed in the literature; an exception is an interesting essay by Patrick Suppes entitled "Models of Data." Suppes points out that from a model-theoretical point of view, one must look to a *hierarchy* of set-theoretical entities mediating between the original model of a given theory and the more limited, finitistic models that correspond to the data churned out by the experiment itself.[37]

Historically, a passionate concern for the cultivation of data has been associated with sciences of some theoretical inadequacy. Compare, say, Professor Siegmund Brandt's *Statistical and Computational Methods in Data Analysis* (New York: American Elsevier, 1970) with Professor Henri Theil's *Principles of Econometrics* (New York: John Wiley & Sons, 1971). The first is 263 pages of neatly composed and up-to-date information: random variables, least squares, minimization—the standard budget of statistical and methodological topics. The second is slightly more than 700 pages and contains a stunning amount of technical and demanding material. One can mush on through Brandt's book in a week; Theil's takes a year, even for sophisticated students of statistics. Brandt's book is composed of lectures first delivered to physics students and particle physicists at Heidelberg University; Theil's is intended for the education of the social scientist.

Despite this, it is useful to judge *World Dynamics* against certain standard guideposts in econometrics.[38]

(1) Economic and econometric theory diverge most obviously in the manner in which relationships are specified. Generally, equations of economic theory are in *structural* form; those of econometric theory in *reduced* or *final* form. Suppose that income i is defined as the sum of consumption c and autonomous expenditures e, while consumption is set as a linear function of income. The equations

$$i = c + e, \quad c = \alpha + \beta i, \tag{2.23}$$

are *structural*: the first is an *accounting* identity; the second, a *behavioral* relationship. Actually, consumption should logically be a function of *disposable* income, so that

[37] Patrick Suppes, "Models of Data," in *Studies in the Methodology and Foundations of Science* (New York: Humanities Press, 1969), pp. 24–35.
[38] I follow the sequence suggested by Michael Intriligator's "Econometrics and Economic Forecasting," in J. M. English, ed., *Economics of Engineering and Social Systems* (New York: John Wiley & Sons, 1972), pp. 151–167.

$$i = c + e, \qquad c = \alpha + \beta(i - t), \tag{2.24}$$

where t = taxes.

The two equations in (2.24) comprise a theoretical statement, however brief, about economic behavior. It is traditional in economics to
congregate variables into *exogenous* and *endogenous* groups, with the
values of the first fixed in, and the values of the second determined *by*,
the economic theory. Endogenous variables above are i and c; exogenous variables, e and t. What should be important in any forecasting
model is the conversion of (2.24) into a structure plotting endogenous
against exogenous variables alone (or against exogenous variables together with *lagged* endogenous variables). Such is the *reduced form* of
the structural equations. A simple rearrangement of (2.24), for example,
yields

$$i = \frac{\alpha - \beta t + e}{1 - \beta}, \qquad c = \frac{\alpha - \beta t + \beta e}{1 - \beta}. \tag{2.25}$$

A further rearrangement of such equations purging the right-hand sides
of any lagged endogenous variables gives the *final form* of the structural
equations.

(2) Differences between structural and reduced equations are a matter
of *form*: the conversion of the one to the other usually reflects the passage
from economic to econometric theory. But equations in final form are
not fit for estimation: variables are often missing, relationships mis-
specified, parameters misestimated. A standard tactic in the face of uncertainty is to add a *stochastic-error* term to equations not strictly accounting identities. For example, one might write

$$c = \alpha + \beta i + \mu, \tag{2.26}$$

where μ is a *random variable*—a number taken once from a set of numbers obeying a fixed probability distribution, usually Gaussian.

(3) Econometric data are inevitably aggregated, either via bunched time
series or clumped cross sections. Aggregation itself gives rise to special
problems, and a separate section is reserved for these below. But other
problems arise from the play between the facts and the theory invoked
to explain them: multicollinearity, serial correlation, structural change,

identification.[39] What remains when these are cleared away are issues that pertain purely to the fitting of data. Here *regression analysis* holds sway. A simple and widely used technique is *least-squares estimation.* In the equation (2.26), for example, least-squares estimators of the parameters are obtained from the data $[(i_1, c_1), (i_2, c_2), \ldots, (i_T, c_T)]$ by minimizing

$$S = \sum_{t=1}^{T} e_T^2 = \sum_{t=1}^{T} [c_t - (\alpha + \beta i_t)]^2 \tag{2.27}$$

with respect to α and β.

The Gauss-Markov theorem has a certain stabilizing force in the application of least-squares techniques. So long as the underlying theory is essentially linear, the theorem guarantees estimates to be (1) linear transformations of the sample data, (2) unbiased, and (3) minimally variant. This points up the profound role of linearity assumptions in econometric theory; of importance, too, is the transformation of structural to final form, since often only the final form of the theory meets the assumptions of the theorem.

(4) An estimated model is not yet a perfect vehicle either for the analysis of structural change or for purposes of forecasting. Determining *comparative statics* involves detailed mechanical manipulations; and projecting a model usually requires the creation of a *forecast horizon* in which specific predictions are bounded by *confidence intervals.* A specific estimation of x at some time t, for example, is usually derived by analysis of a probability distribution; x_t^{est} will be the expected mean value of the distribution determined by the smear of data culled before t. A 95 percent confidence interval, then, would be a range of values about x_t^{est} such that the probability of a *measured* value of x falling within the interval is 0.95. Such confidence intervals inevitably fan out as $t \to \infty$, often very rapidly. Thus sticking to 95 percent as the desired probability sooner or later entails a gargantuan interval; conversely, sticking to a fixed interval drives the probability downward.

To survey some of the standard issues that arise in the conversion of

---

[39] See F. M. Fisher, *The Identification Problem in Econometrics* (New York: Mc-Graw-Hill, 1966). Philosophers will recognize the identification problem as the econometric correlate to the problem of radical translation; see W. V. Quine, *Word and Object* (Cambridge, Mass.: The MIT Press, 1961).

economic to econometric theory is to acquire a certain sober sense for the failings of systems dynamics. Taking the points I have glossed in order: there is no cultivation in *World Dynamics* of equational structure that goes much beyond simple numbering; no deference to stochastic estimation; nothing by way of careful estimation of economic or political data by means of regression analysis; no sense, finally, that confidence intervals for even the most ambitious of econometric theories inevitably decay after a few months at most (the theories of *World Dynamics* and *The Limits to Growth* run on to the next century). Moreover, there is nothing in *World Dynamics* having to do with the careful estimation of time series or aggregated data generally, and there are no standards of goodness of fit. At every point crying for methodological sophistication, Forrester and Meadows leap to the cannon's mouth with theories of superb simplicity. Neglected is not simply econometric esoterica, though that is missing too, but the entire body of statistical technique that is generally demanded by quantitative work in the social sciences.[40]

Sociometric *schweinerei* might be offset by sparkling theoretical insights: there is economics without econometrics. But while the applications are unsophisticated, the theory is unsatisfactory, so *World Dynamics* and *The Limits to Growth* get clobbered coming and going.

On Lawlikeness

Any idiot can draw a graph.

Nelson Norgood[41]

The aim of science, grandly conceived, is to describe a realm of the lawlike existing beneath the flux and fleen of things. To the theory belong the laws, but econometricians sometimes write as if their models were objects of occasional accuracy having nothing to do with that system of relationships largely the province of classical economic theory. It is thus instructive to note that many large-scale econometric models are essentially statistical elaborations of Keynesian theory. In an interesting article, for example, Lawrence Klein describes a para-

---

[40] Professors Forrester and Meadows have included *sensitivity tests* in the description of their models. Generally, these involve quadrupling some parameter and then noting the undeflected arrival of catastrophe on schedule or some years later. This is a procedure about as reliable as buying several copies of the morning newspaper in order to assure yourself that what you read in the first is true.

[41] Nelson Norgood, *Sociological Sociometrics* (Chicago: The Inverse Press, 1969), p. 579.

metric, linear version of the Lange-Hicks equations and then shows what by way of additional assumption is needed to convert this essentially static and unspecified system into a full-fledged if rather modest econometric theory. What results is nonetheless a Keynesian theory, even though the Lange-Hicks equations can be expressed in four sentences while the econometric model requires thirteen items ranging over five exogenous and thirteen endogenous variables. "We call this a Keynesian model," Klein writes, "even though it looks vastly different from the small four-equation Keynesian system presented earlier, because if prices are given it determines effective demand."[42]

The role of theory in the creation of forecasting models is quite central: without some compelling assumptions about the nature of economic relations, what one has to go on are data drawn from a historical record and projected forward by techniques that assume some simple continuity in the behavior of economic agents.

An econometric model must track the data in a natural and intuitively plausible way, one that brings out a sort of congruence between the qualitative structure of the theory and the domain of relations it is meant to explain. The Klein-Goldberger model of the United States economy consists of twenty-five difference equations in as many endogenous variables. Its qualitative properties were not perfectly understood when it was first presented; analytical appraisals were restricted to artificial cases. Then the Professors Adelman simulated the system and ran it forward into the future to learn whether the model depicted the oscillations of a modern industrial economy, i.e., the so-called business cycles. To the question of "what sort of time paths [the] equations generate in the absence of additional constraints or shocks," the Adelmans answered unequivocally that the system was linear, even monotonic, and that far from looping in cycles, it surged ever forward. So much the worse for the Klein-Goldberger model: a gap of such qualitative intractability between theory and reality is evidence, if anything is, that a model is simply wrong. But the Adelmans raised a redeeming possibility: when subjected to exogenous shocks, the system *did* oscillate in a way that matched the swings of the American economy. This suggests that a plausible model of the United States economy might be the Klein-Goldberger model randomly perturbed.

That oscillatory phenomena can be depicted by shocking linear sys-

[42] Lawrence Klein, "What Kind of Macroeconomic Model for Developing Economies?," *Econometric Annual of the Indian Economic Journal*, Vol. 13, No. 3 (1965), pp. 313–324.

tems has been known, of course, at least since Frisch and Slutsky raised the possibility of a new class of models for business cycles. But from the point of view of the qualitative satisfactoriness of the Klein-Goldberger model itself, two quite different conclusions may be drawn from the Adelmans' experiment:

On the one hand, if one wishes to retain the hypothesis that periodic cumulative movements are self-generated in the course of the growth process in a realistic economy, one may contend that the Klein-Goldberger model is fundamentally inadequate, and hence that it is inapplicable to further business cycle theory. On the other hand, one may hold that, to the extent that the behavior of this system constitutes a valid qualitative approximation to that of modern capitalist society, the observed solutions of the Klein-Goldberger equations imply that one must look elsewhere for the origins of business fluctuations.[43]

Although the two choices staked out here are flatly incompatible, there is nothing, so far as I can judge, that suggests an empirical difference. The dispute, rather, is between two qualitatively distinct views of the nature of capitalist economies both of which can be reconciled with the existing data.

Is population growth lawlike?

What of the chief dynamical assumptions to *World Dynamics*? The governing assumptions to the theory, it is prudent to remember, describe as exponential the rise of population, capital formation, and the generation of pollution. This introduces an unavoidable infelicity: in the very nature of things exponential growth cannot be invoked to explain movement toward fixed *limits*. What is needed to account for the *whole* of the data pertaining, say, to population growth is an expression describing a curve only part of which indicates very rapid growth.

   A simple model illustrating this point is created when an addition is made to the blank assumption of exponential growth: postulate a maximum population $M > 0$ sustainable by a given environment, together with a minimum population $0 < m < M$ below which a species will die out, and assume that the growth of a population is proportional not only to x, the present population, but to $M - x$. A natural equation describing the growth of population under such conditions is

$$\frac{dx}{dt} = c(M - x)(x - m)x, \tag{2.28}$$

[43] Adelman and Adelman, "The Dynamical Properties of the Klein-Goldberger Model."

where c is a constant greater than zero. This is an equation with three equilibrium solutions: $x = M, 0,$ or m. The first two are stable, while the third is not.[44]

Now (2.28), crude though it unquestionably is, suggests a dynamical interpretation of population growth distinctly unlike that offered by Forrester. Populations growing at some fixed rate, according to (2.28), will achieve a maximum and then, since M represents a stable equilibrium, simply stay there.

Classical studies of the dynamics of population—to take a more sophisticated example—usually begin with the Pearl-Verhulst logistic equation:

$$\frac{dx}{dt} = bx \, \frac{M - x}{M}, \qquad\qquad (2.29)$$

where b is a constant representing roughly (neglecting the last term) the increase per organism per unit of time. The term $(M - x)/M$ is thus a "density corrective," indicating that growth slows as the population approaches its biological maximum.

Equation (2.29) also embodies a lawlike hypothesis about the growth of biological populations; it is, no doubt, too general, since useful results require much finer detail, but it does soundly limn the dynamics of growth. In using (2.29), demographers assume that there exist fixed limits to growth, and that rates of growth will vary continuously as the population approaches those limits. This is reasonable and reflects the commonsensical observation that as the size of a population increases, the availability of food, water, and other scarce resources is bound to decrease; consequently, the rate of increase should begin to slow. The graph of the solution looks roughly like a flattened S, indicating very rapid initial growth followed by an asymptotic approach to a fixed limit.[45]

[44] See Plaat, *Ordinary Differential Equations*, pp. 62-63. This is an excellent elementary introduction to the basic theory of ordinary differential equations.
[45] The classic texts for the mathematical analysis of population growth are Raymond Pearl, *Introduction to Medical Biometry and Statics* (3rd ed., Philadelphia: Saunders), and A. J. Lotka, *Théorie analytique des associations biologiques*, Part II, *Analyse démographique avec applications particulière à l'espèce humaine* (Paris: Hermann et Cie.). Lotka's work especially has a mathematical sophistication of a sort that is absent from the writings of Meadows and Forrester. For the basic information about worldwide population trends, see *The Determinants and Consequences of Population Trends*, Vol. 1 of the United Nations Population Studies, No. 50.

The *whole* of the theory of *World Dynamics* does not suggest a population curve ascending exponentially toward infinity. But the dynamical equations taken at their "normal" value (with multipliers all set to unity) do describe *purely* exponential processes, as we have seen in (2.12). Thus, if they are to yield realistic curves, the equations that actually figure in the system (from 1900 on) require multipliers that can produce perturbations or shocks, as in (2.14).

Talk of perturbations is meant to suggest the example provided by the Klein-Goldberger model. The numerical results and even the geometrical character of Forrester's theory may accord with the facts, but its cardinal theoretical processes are realistic only under occult influences. What the analysis gains in realism it loses in intuitive plausibility, for the variables or functions that act to change exponential growth, while they are endogenous to the theory as a whole of *World Dynamics*, are exogenous to the theory of population growth. Comparing curves generated by the Pearl-Verhulst logistic equation and curves generated by the theory of *World Dynamics* may reveal no quantitative discrepancies, but only the former, it seems to me, is an appropriately lawlike object, an intrinsically satisfying and qualitatively accurate accounting of the mechanism by which populations grow. What is missing in the latter is some sense of the movement of population growth as a single phenomenon subject to lawful analysis, like the evaporation of perfect gases or the consumption of tea in a perfectly competitive economy.

There may very well be *no* lawful account of the mechanism of population growth: the careful researcher may have to forego the lawlike and tote up individually those forces that affect population growth, much as the engineer can appeal to no *specific* law of motion that will describe the behavior of a spaceship as it passes, say, through wobbles due to engine failure. But nothing suggests that population growth is peculiarly exempt from lawlike analysis, that it is a phenomenon too complex and too idiosyncratic to sustain a single lawlike explanation.[46]

The pivot on which traditional population theory turns is the connection between *density* and *fecundity*, a connection the logistic equa-

---

[46] For an account of the conceptual difficulties involved in fitting a system of differential (or integral) equations to the empirical data relating fertility, fecundity, and density, see Kenneth Watt's *Ecology and Resource Management* (New York: McGraw-Hill, 1968), Chapter II, section 11.4. A good introduction, on an elementary level, to some of the detailed methods of demography is N. Keyfitz and W. Flieger, *Population: Facts and Methods of Demography* (San Francisco: W. H. Freeman, 1971).

tion exploits in mapping increases in population inversely against increases in density. It is this hinge, too, that the theory of *World Dynamics tacitly* invokes: density and population size are linked indirectly through the effect the first has on the stock of natural resources and thereby on the rate of population growth itself. But it is a connection never exploited in *World Dynamics*; where there should be a simple relationship yielding the particular curves of *World Dynamics* as special cases, analogues to those elegant curves describing the growth and decline of fish populations in landlocked streams, there is nothing at all.

World 3, the model featured in *The Limits to Growth*, evidently improves on this aspect of World 2, for population growth is there explained on the assumption that the structure of cause and effect regulating the growth of biological populations is lagged in a way that suggests a common pattern of overshoot and collapse.

Biologists understand these relationships.[47] Assume, for example, that the rate of increase of a given population is a function of its density, with the rate slowing gradually as density increases. This is simply the assumption behind the Pearl-Verhulst equation. (Expressing rates of increase as functions of population density is an elegant way of recording such indirect forces as crowding, food supply, and the spread of disease that can affect, directly or indirectly, the nature of a population's density.) Now it seems natural to suppose that fertility is in some sense a *lagged* function of density. On the assumption that an organism will register the effects of a continuous increase in population density in a sort of weighted average over the period of its prereproductive years, one would expect, as population density rises, that organisms would overbreed for conditions of density which they are in fact experiencing. Given an absolutely fixed limit to growth and the symmetrical overshooting that occurs on the downward swing, the resulting oscillations might be described by the lagged equation

$$\frac{dx}{dt} = bx_{t-t_1} \frac{M - x_{t-t_2}}{M},$$
(2.30)

where t_1 represents the time required to commence reproduction when conditions are favorable and t_2 is the time required to register changes in density.

A difficulty with this suggestion is that the equation postulates a

[47] See P. J. Wanersky and W. J. Cunningham, "On Time Lags in Equations of Growth," *Proceedings of the National Academy of Sciences*, Vol. 42 (1956), pp. 699-702.

linear relationship between fertility (or fecundity) and density; plotting b, the increase per organism per unit of time, against x, the number of individuals alive at a given time, should yield a straight line. But there is overwhelming evidence that in fact, as one would expect, the relationship between fertility and density is highly nonlinear. An obvious condition on any theoretical model, Kenneth Watts suggests, is that appropriate curves tracing such relationships as those that hold between fecundity and density be easily derivable. This condition fails for the lagged model at (2.30), and it will be interesting to see if it holds for the detailed theories of population dynamics promised by Meadows and associates.

Quality and quantity
Consider the Volterra-Lotka predator-prey equations.

Two species x and y are such that in the usual course of things x preys on y—wolves on deer, for example.[48] Suppose that were there no wolves, the deer, whose source of food is fixed, would grow according to the relationship

$$\frac{dy}{dt} = \alpha y, \tag{2.31}$$

where α is constant. Wolves feed *only* on deer, so that if $y = 0$, the wolves die off:

$$\frac{dx}{dt} = -\delta x. \tag{2.32}$$

Assume now that $xy > 0$. The fraction of devoured deer should be roughly proportional to the number of wolves, so that the deer will vary according to

$$\frac{dy}{dt} = (\alpha - \beta x)y, \qquad \beta > 0. \tag{2.33}$$

By similar reasoning, the fraction of surviving wolves will be proportional to the number of deer,

---

[48] I follow Hirsch and Smale here, but the material is standard. See M. W. Hirsch and S. Smale, *Differential Equations, Dynamical Systems and Linear Algebra* (New York: Academic Press, 1974), pp. 258–263.

$$\frac{dx}{dt} = (\gamma y - \delta)x, \qquad \gamma > 0, \tag{2.34}$$

since $\gamma y x$ comprises a *reduction* in the death rate.

Taken together, (2.33) and (2.34) constitute the Volterra-Lotka predator-prey equations. They are usually investigated on the XY plane; instead of considering solutions as curves in three-dimensional XYT space, we stick to projections of these curves onto the XY plane.

The Volterra-Lotka equations have two singular points: the origin and $(\alpha/\beta, \gamma/\delta)$. There are also two obvious trajectories: $x \equiv 0$, $y = y_0 e^{\alpha t}$; $y \equiv 0$, $x = x_0 e^{-\delta t}$. These are the only explicit solutions, and the problem, typical in the context of the qualitative theory of ordinary differential equations, is to come up with a global analysis without actually solving the equations.

The phase portrait of the equations' solutions may be investigated by sectioning the plane into quadrants determined by the lines $x = \alpha/\beta$ and $y = \delta/\gamma$, as in Figure 2.5. The point $(\alpha/\beta, \gamma/\delta)$ is singular: trajectories around it can form a closed orbit, spiral toward the origin, or spiral toward the point itself. The phase portrait must give this information. Inspection of the equations reveals that solutions will move in a counterclockwise direction around the singular point.

As it happens, the predator-prey equations are amenable to fairly straightforward investigation. The essential trick involves obtaining an *integral* for the equations, that is, a function of the form

$$H(x, y) = F(x) + G(y), \tag{2.35}$$

where solution trajectories are level curves of $H(x, y)$. For systems with an integral, a complete description of the phase space may be obtained from consideration of the curves $H(x, y) = $ constant. An appropriate H in the present instance is

$$H(x, y) = \beta x - \alpha \log x + \gamma y - \delta \log y, \qquad xy > 0, \tag{2.36}$$

from which we infer that:
1. The singular point is stable since H is a Liapunov function.
2. There are no limit cycles since H is not constant on an open set.
3. Every trajectory is closed, excepting of course the two point trajectories.

The phase portrait therefore looks like Figure 2.6.

While the equations that make up the Volterra-Lotka model of

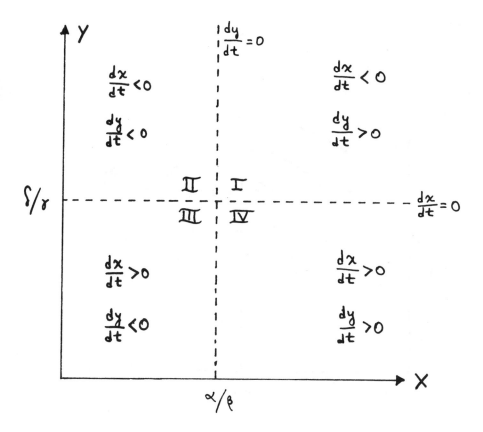

Figure 2.5
A sectioning of the plane useful for examining the phase portrait of solutions to
the predator-prey equations.

predator-prey interaction are simple, the assumptions that govern
them are ruthless and their analysis subtle. This suggests a maxim for
the mathematical modeler: start simply and use to the fullest the
resources of theory. It is this prescription taken neatly in reverse that
characterizes *World Dynamics* and *The Limits to Growth*; there the
prevailing pedagogical maxim has been: pile up an imposingly complex system of equations and then subject them to an analysis of
ineffable innocence. It is a natural prescription for theorists ignorant
of differential theory.

Here is another lesson afforded by the predator-prey equations:
it is not always necessary to subject an analytically intractable system
to simulation in order to understand it qualitatively; correspondingly,
qualitative insights are at greater depth than partially quantitative results. The moral: look to systems for which a qualitative analysis is

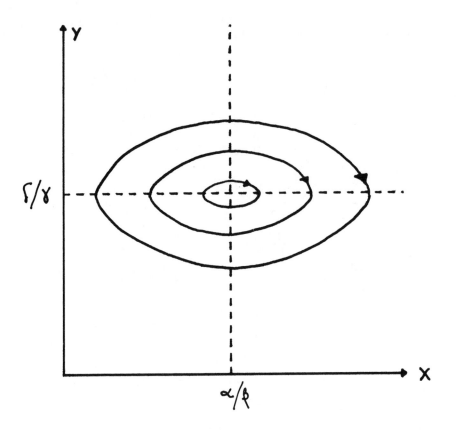

Figure 2.6
Phase portrait of solutions to the predator-prey equations.

possible. Although nothing is known of explicit solutions to the predator-prey equations, except for the trivial case, everything relevant is
known about the *general class* of solutions: they are all periodic about
a stable center.

It is worth noting, too, that the progression of qualitative insights
that lead up to the final predator-prey equations differs markedly
from the insights Forrester and Meadows record in informal accounts
of their theory. *There* one has a two-step process: exponential growth
modulated by the multipliers. The predator-prey equations do not
admit of an artificial distinction between processes understood inherently and the same processes tamed by coefficients; the coefficients are all defined generally. Moreover, the interaction between
dynamic variables, which in turn affects the coefficients, is expressed
by the dynamical equations themselves, so that there is no need here
for graphs, look-up tables, or charts.

Finally, the prevailing mathematical modesty that informs the predator-prey equations is worth an admiring salute. Successful science is abstractive: of all that is of interest, only a fragment is susceptible to explanation, and only a fragment of that fragment to serious mathematical analysis. To the zealot, an ecological community might suggest vast opportunities for mathematical modeling—an equation for each species perhaps, resulting in a system sufficient to bankrupt core storage. Power and scope, however, are often *inversely* related; while the predator-prey equations offer an appraisal of reality that is at once severely idealized and incomplete, what insights they offer are at some depth. In *World Dynamics* and *The Limits to Growth* the reverse holds true: the scope is unbounded but the insights are slight.

Determinism

Given the equations of *World Dynamics*, subsequent states of the system occur with the implacability of the rotation of Saturn. But human affairs require the powers of the calculus of probability. The various "dynamics" are distinctly unprobabilistic, but this disadvantage the econometrician can usually alleviate by the inclusion of stochastic variables. Yet the aggregates of *World Dynamics* suggest something more radical: *a fundamentally probabilistic structure*, with the theory as a whole shaped to reflect the stochastic nature of social life. A vast body of operations research has been directed to such theories. One thinks here of random walks and Markov processes— stochastic phenomena in which change occurs at fixed intervals; there are systems of equations, such as those required to model the so-called pure birth process, that blend differential and stochastic features; obvious examples are the Chapman-Kolmogorov equations, the Forward equations, and the Fokker-Planck equations. The intuitively attractive machinery of the theory of ordinary differential equations adapts quite easily to the stochastic case: what results are *stochastic differential equations* which, despite their kinship to the deterministic case, often yield significantly different results. In the theory of epidemics, for example, deterministic models yield damped oscillations and corresponding stochastic models do not.[49]

I have no argument that global theories *must* turn to nondeterministic equations; still, there is something like a generalized argument from experience. Probabilistic models have been exceptionally useful

[49] See Norman T. Bailey, *The Mathematical Approach to Biology and Medicine* (New York: John Wiley & Sons, 1967), Chapter 9, and M. S. Bartlett and R. W. Hiorns, eds., *The Mathematical Theory of the Dynamics of Biological Populations* (New York: Academic Press, 1973), Chapter 2.

in the analysis of systems whose chief features are not qualitatively clear; for example, mathematical geneticists have recently used probabilistic diffusion processes as models for genetic change. Having hit on differential equations as the instrument by which social change is recorded, Forrester and Meadows might have moved toward the broad class of stochastic differential equations. They would have found it a flexible tool.

Aggregation

The entities over which a theory ranges are, from the perspective of the theory itself, its basic stock.[50] Cell biologists study cells; molecular biologists, the macromolecules that make them up. Microeconomics surveys a domain of firms and individuals; macroeconomics, an extended domain of national income, gross national product, and national demand. The *inner* perspective of the theory corresponds to a view of the world determined *by* the theory, but theories are generally set in a larger intellectual perspective; and so even if a given theory is restricted to a class of entities about whose structure it says nothing whatsoever—solid-state physics, for example—theoreticians will experience a natural curiosity about the entities that *compose* the entities they invoke.

The distinction between the structure and the surface of things corresponds to a switch in levels. Microtheoretical entities are the components of their bulkier macrotheoretical counterparts, but this is only apparent once the narrower perspectives of the microtheory have been breached. From the transcendental viewpoint, taking in structure and surface alike, the distinction between micro- and macrotheories is general throughout science. There is thermodynamics, a macroscopic theory, and statistical mechanics, the systematic view of structure to which it can be reduced; geometrical optics and the quantum theory of light; many-body physics and the theory of elementary particles; the hydrodynamic theory of the field and the global theory of weather prediction. In mathematics, the distinction between macro- and microtheory has an analogue in the familiar distinction between differential topology and global analysis.

Chemistry, the chemist often notices, can usually be conducted without thought of the reduction of chemical constituents to their molec-

---

[50] See P. Achinstein, "Macro-Theories and Micro-Theories," in Patrick Suppes et al., *Fourth International Congress for Logic, Methodology, and Philosophy of Science* (Amsterdam: North Holland, 1973), pp. 533–567, for an interesting if incomplete discussion.

ular or atomic structures; so can solid-state physics, cell biology, and the theory of Fermi surfaces. Mach, in fact, argued that chemistry had *no* use for atomic theory; his was a view of the sciences marked by a stern segregation of levels. In the social sciences and economics, however, the situation is quite different. The very entities over which macroeconomics ranges—aggregate demand, gross national product, national income—require definition in terms drawn from microscopic theories; much the same is true for ecological analysis in the political sciences. Macroeconomics and political science are made up of inaccessible aggregates that must be defined or otherwise made intellectually palpable before they can be studied. No doubt this is what R. G. D. Allen had in mind when he wrote that a purely macroeconomic theory cannot be "satisfactory to an economist" since the macroeconomic relations are "derived constructions."[51]

World Dynamics and *The Limits to Growth* are highly aggregated theories, as are the models discussed in *Industrial Dynamics* and *Urban Dynamics*. In form they are macropolitical, analogues to macroeconomics, but the scale of aggregation is so vast as to suggest a Hegelian passage from quantitative to qualitative change. Such are theories that trade on elevated aggregation, a condition which resembles elevated blood serum cholesterol in overall effect.

Questions of economic aggregation were first raised distinctly by Lawrence Klein. "Many of the newly constructed mathematical models of economic systems," he wrote,

especially the business cycle theories, are very loosely related to the behavior of individual households or firms which must form the basis of all theories of economic behavior. In these mathematical models, the demand equations for factors of production in the economy as a whole are derived from the assumption that entrepreneurs collectively attempt to maximize some aggregate profit; whereas the usually accepted assumption is that the individual firm attempts to maximize its own profit. For example, Evans, Hicks, Keynes and Pigou all have

[51] R. G. D. Allen, *Mathematical Economics* (New York: St. Martin's Press, 1956), p. 694. Inaccessible aggregates are not entirely the rule in the social sciences. Paul Lazerfeld and Herbert Menzel have drawn a distinction between "global" and "analytic" variables: by the latter they mean the inaccessible variables, the creations of theory; by the former, the truly autonomous macroscopic variables that correspond in a rough way to variables of state in thermodynamics, such as pressure and heat (see P. Lazerfeld and H. Menzel, "On the Relationship Between Individual and Collective Properties," in Amitai Etzioni, ed., *Complex Organizations* [New York: Holt, Rinehart & Winston, 1965]). M. T. Hannon remarks quite accurately, however, that global variables are in short supply in the social sciences (see M. T. Hannon, "Problems of Aggregation," in H. M. Blalock, Jr., ed., *Causal Models in the Social Sciences* [Chicago: Aldine Publishing Co., 1971]).

in their systems marginal productivity equations for the total economy. . . . These marginal productivity equations are written, without
justification, for the economy as a whole in exactly the same form as
the marginal productivity equations for a single firm producing a single
commodity.[52]

This suggests a view of aggregation dominated by the difficulties involved in *projecting* a macroscopic theory from a microscopic theory
given only the latter together with a set of aggregating relationships.
But, equally, one might think of the problem in terms of macro- and
microscopic theories, with the cardinal issue the *existence* of suitable
systems of aggregation between them. Subsequent literature has split
along these lines.[53]

The problem, then, is to set out conditions on aggregation, and Klein
proposed two. First, functional relationships defined for individuals
must also be defined for aggregates: if there is a production function
that connects outputs to inputs at the *i*th firm, there must also be an
overall production function that matches aggregate outputs to aggregate
inputs, though its mathematical properties may be quite distinct from
those of any of the individual production functions. Second, if profits
are maximized by individual firms so that marginal productivity equations hold under conditions of perfect competition, then aggregate
variables must also be embedded in aggregate or macroscopic equations
satisfying the same laws of marginal productivity.

But under what general conditions will these criteria be met? This is
the econometrician's *pure problem of aggregation* ("pure" in contrast
to problems that arise when perfect aggregation is acknowledged as
impossible and the goal instead involves minimizing *aggregation bias*).

In the simplest case

$$y = F(X), \tag{2.37}$$

where F is, say, a *production function* and X is the set of variables
$\{x_1, x_2, \ldots, x_n\}$ upon which F depends. Generally F will be complicated and n large; obviously some simplification ensues from

---

[52] L. R. Klein, "Macroeconomics and the Theory of Rational Behavior," *Econometrica*, Vol. 14 (April 1946), pp. 93–108.

[53] See H. Theil, *Linear Aggregation of Economic Relations* (Amsterdam: North
Holland, 1954); K. O. May, "The Aggregation Problem for a One Industry Model,"
Econometrica, Vol. 14 (October 1946), pp. 285–298; H. A. J. Green, *Aggregation
in Economic Analysis* (Princeton, N.J.: Princeton University Press, 1964). Green's
monograph is the most complete technical account of aggregational difficulties in
the literature.

a regrouping of variables to take advantage of any functions that mediate between the initial and the final products: for example,

$$F(X) = G[H(x_1, x_2, \ldots, x_k), J(x_{k+1}, x_{k+2}, \ldots, x_n)] \qquad (2.38)$$

amalgamates variables and substitutes for them the values of intermediate functions. Under what conditions can we find such functions? The answer was provided by Wassily Leontieff, who argued roughly that the condition for *functional separability* was just that the marginal rate of substitution for any variable within a subgroup of variables—those clumped under an intermediate function—depend only on the variables *within* that subgroup.[54]

Robert Solow has provided a simple example.[55] Consider a production function

$$Q = F(L, C_1, C_2), \qquad (2.39)$$

where Q is a single output, L is an input of a single grade of labor, and C_1 and C_2 are inputs of services of two distinct kinds of capital equipment. We can achieve a separation

$$Q = F(L, C_1, C_2) = H(L, K),$$
$$K = J(C_1, C_2), \qquad (2.40)$$

only when—and this is Leontieff's chief result—

$$\frac{\mathrm{MPP}_1}{\mathrm{MPP}_2} = \frac{\partial F/\partial C_1}{\partial F/\partial C_2} \equiv \frac{\partial H/\partial K \cdot \partial J/\partial C_1}{\partial H/\partial K \cdot \partial J/\partial C_2} \equiv \frac{\partial J/\partial C_1}{\partial J/\partial C_2}, \qquad (2.41)$$

where the last expression is independent of L. Here MPP_1 denotes "marginal physical productivity"; Leontieff shows that the *ratio* of MPP_1 to MPP_2—the marginal rate of substitution of C_1 for C_2—must be fixed independently of the values of L in order for (2.39) to pass to (2.40).

This is natural but cramping; and the point goes beyond the purely econometric case since the Leontieff-Sono result is a theorem in (ele-

[54] Wassily Leontieff, "Introduction to a Theory of the Internal Structure of Functional Relationships," in *Essays in Economics* (New York: Oxford University Press, 1966).
[55] Robert Solow, "The Production Function and the Theory of Capital," *Review of Economic Studies*, Vol. 23, No. 2 (1955–1956), pp. 101–108.

mentary) functional analysis. For highly aggregated theories, such as those adumbrated by *World Dynamics* and *The Limits to Growth*, the theory's functions *collapse*: a complete representation would require endless detail. The multipliers, for example, are ultimately functions of the system's states; the states, in turn, are aggregate entities, and were the multipliers pegged to unaggregated indices, they would no doubt turn out to be infinitely more complicated than they are.[56] Now we must ask whether the expressions that make up the chief functions of the theory stand toward some hypothetically unaggregated function as (2.40) stands to (2.39). Given the nature of the relationships that the aggregate theory itself suggests, it seems highly unlikely that the original stock of variables would be functionally separable. This, of course, is what one expects from the Leontieff result: the conditions for functional separability are stringent. But this raises the possibility that there may be no microscopic theory, no disaggregated set of functions, that can be related to the macroscopic theory. And vice versa: for such disaggregated theories as we may happen upon, the high degree of interrelationship between variables may not make uniformly consistent aggregation to a macroscopic theory possible.

Elevated aggregation does not affect only the intermediate products in a production function. The multipliers embody aggregated *relations*: the downward-sloping curve of extractive inefficiency in Figure 2.1 is meant to reflect the experience of many separate industries banded together into one curve of functional dependence.

In the abstract, then, we start with m individual functions

$$y_j = f_j(x_{1j}, x_{2j}, \ldots, x_{nj}), \quad j = 1, 2, \ldots, m, \tag{2.42}$$

with all $x_{ij} \geq 0$.[57] What we would like to do is convert (2.42) to

$$y = F(x_1, x_2, \ldots, x_n), \tag{2.43}$$

where y and the x_i are aggregating functions,

---

[56] The technical version to *The Limits to Growth* evidently makes mention of a production function of the "Walrus-Leontieff-Harrod-Domar type," an object of considerable grandeur. But there is no indication that this function is, in fact, functionally separable. See C. M. Cooper's Appendix to Chapter 6 of Cole et al., *Models of Doom.*

[57] Here I follow Green's monograph. See his Chapter 5, pp. 35–38, for further details.

$$y = y(y_1, y_2, \ldots, y_m),$$

$$x_i = x_i(x_{i1}, x_{i2}, \ldots, x_{im}), \qquad i = 1, 2, \ldots, n. \tag{2.44}$$

Clearly a *necessary* condition for consistent aggregation is that for all i and j

$$\frac{\partial F}{\partial x_i} \frac{\partial x_i}{\partial x_{ij}} = \frac{\partial y}{\partial y_j} \cdot \frac{\partial f_j}{\partial x_{ij}}; \tag{2.45}$$

i.e., marginal properties to consume must all be constant and equal. (This is just Klein's second criterion, on the assumption, of course, that the equations still range over *economic* relations.)

But according to a theorem by Nataf,[58] (2.45) implies that (2.42) can be consistently aggregated to (2.43) if and only if there exist functions $G, H, g_i, h_j, G_i, H_j, g_{ij}, h_{ij}$ such that

$$y = H[h_1(y_1) + h_2(y_2) + \ldots + h_m(y_m)] \tag{2.46}$$

$$= G[g_1(x_1) + g_2(x_2) + \ldots + g_n(x_n)],$$

where

$$y_j = H_j[h_{1j}(x_{1j}) + h_{2j}(x_{2j}) + \ldots + h_{nj}(x_{nj})], \qquad j = 1, 2, \ldots, m, \tag{2.47}$$

and

$$x_i = G_i[g_{i1}(x_{i1}) + g_{i2}(x_{i2}) + \ldots + g_{im}(x_{im})], \qquad i = 1, 2, \ldots, n. \tag{2.48}$$

And this is an impressively narrow conclusion showing, in Green's words, "the remarkable importance, in aggregation problems, of the *linearity* of the functions involved, be they micro-relations, macro-relations, or functions defining aggregates."[59]

[58] A. Nataf, "Sur la possibilité de construction de certains macro-modèles," *Econometrica*, Vol. 16 (July 1948), pp. 232–244.
[59] Green, *Aggregation in Economic Analysis*, p. 36; emphasis added.

For example, consider the case in which aggregates are defined as simple sums:

$$y = \sum_{j=1}^{m} y_j; \quad x_i = \sum_{j=1}^{m} x_{ij}, \quad i = 1, 2, \ldots, n. \quad (2.49)$$

By (2.45),

$$\frac{\partial F}{\partial x_i} = \frac{\partial f_j}{\partial x_{ij}}. \quad (2.50)$$

Since $\partial F/\partial x_i$ depends *only* on the total x_i, the values of $\partial f_j/\partial x_{ij}$ must be equal for all j and constant for all x_{ij}. So individual functions must be linear with identical slopes:

$$y_j = a_j + \sum_{i=1}^{n} b_i x_{ij}. \quad (2.51)$$

With the goal of consistent aggregation in mind—what Theil called "perfect aggregation"—the theorems of Leontieff and Nataf, obvious though they are, must be reckoned depressing. This is clear in the economic case: Leontieff's results have the effect of paring away innumerable functional relationships as deficient precisely in their ability to collapse on their internal variables, and Nataf's theorem has the unhappy effect of raising *linearity* into a quasi-theological principle. This bodes ill for global theories such as those espoused in *World Dynamics* and *The Limits to Growth*, for these are systems that see in the nonlinear not simply something to be endured but a condition of man and nature to be exalted, like the medieval stigmata.[60]

[60] For other discussions, pitched more to political science, see J. V. Gillespie and B. A. Nesvold, eds., *Macro-Quantitative Analysis* (Beverly Hills, Calif.: Sage Publications, 1970), and M. Dogan and S. Rokkan, eds., *Quantitative Ecological Analysis in the Social Sciences* (Cambridge, Mass.: The MIT Press, 1969). The analogue in political science to Klein's *Econometrica* article is W. Robinson's "Ecological Correlation and the Behavior of Individuals," *American Sociological Review* (June 15, 1951), pp. 351–357. Practical problems of aggregation, of course, have been a trigger for the development of disaggregated theories detailed enough to cover a sizable chunk of the economic actors in a given economy. Such are the microanalytic theories; see, for example, G. H. Orcutt, "Micro-Analytic Models of the United States Economy: Needs and Development," *American Economic Review* (May 1962), pp. 229–240.

Well-Posed Problems in Analysis

Throughout nature we observe continuous changes giving rise to discontinuous jumps. . . . People suddenly change their minds . . . nations suddenly go to war . . . bridges suddenly break and trees suddenly tumble.

E. C. Zeeman[61]

The French mathematician Hadamard, in fashioning the concept of a well-posed problem in analysis, had in mind a selection from among the class of possible dynamical systems of those with *physical* significance. Clearly a dynamical system must admit of uniquely specified solutions if it is to be useful at all; that the solutions must vary continuously with variations in the initial data is a concession to the frailties of human observation and measurement:

Data in nature cannot possibly be conceived as fixed; the mere process of measuring them invokes small errors. . . . Therefore, a mathematical problem cannot be considered as realistically corresponding to a physical phenomenon unless a variation of the given data in a sufficiently small range leads to an arbitrary small change in the solution.[62]

In standard cases, theorems invoking the Lipschitz condition provide the appropriate assurances. Classical models of well-set problems are drawn from elementary dynamics. For example, if n points of mass m_j $(j = 1, 2, \ldots, n)$ attract each other according to Newton's law, their position coordinates will satisfy a system of $3n$ second-order differential equations of the form

$$\frac{d^2 x_i}{dt^2} = \sum_{j \neq i} \frac{G m_j (x_j - x_i)}{[(x_j - x_i)^2 + (y_j - y_i)^2 + (z_j - z_i)^2]^{3/2}}, \qquad i = 1, 2, \ldots, n$$

(2.52)

(with $2n$ similar equations for the y_i and z_i). Suitable theorems guarantee that the initial positions and velocities of the mass points uniquely determine their subsequent paths; the n-body problem is thus determinate. Continuity proves that the n-body problem is well-posed as well.

[61] E. C. Zeeman, "The Geometry of Catastrophe," *Times Literary Supplement* (December 10, 1971).
[62] R. Courant and D. Hilbert, *Methods of Mathematical Physics* (New York: John Wiley & Sons, 1962), Vol. 2, p. 127. See also Hadamard's classic discussion *Le Problème de Cauchy et les Equations aux Derivées Partielles Linéaires Hyperboliques* (Paris: Hermann et Cie., 1932).

Well-posedness taken in this simple fashion does give a sense of what is to count as a physically acceptable model, but unfortunately the Navier-Stokes equation, the final-value problem for the heat equation, the Cauchy problem for the Laplace equation, and the Dirichlet problem for the wave equation are all improperly posed and physically significant.[63]

Still, man is a boundary-marking animal, and something like well-posedness has been urged on theories as an axiomatic accretion by the topologist René Thom in order to effect the work of segregating the satisfying theories from all the rest. Continuous dependence is actually too weak a condition for what Thom wants. Some systems are well-posed but unstable: the effects of small errors in measurement accumulate unwholesomely because the chief theorems warranting well-posedness hold only for specific intervals (on the line) or regions (in the plane); beyond these intervals, continuously varying but unstable solutions might spread apart like the edges of a fan.

Structural stability is a concept that Thom sees as crucial: in its over-all aspects, it is the most general notion dealing with the question of whether a system of differential equations preserves its qualitative structure in the face of perturbation. In the simplest case,

$$\frac{dx}{dt} = f(t, x), \tag{2.53}$$

a solution $\Phi(t)$ is said to be *stable* if for any $\epsilon > 0$, there is a $\delta > 0$ such that *any* solution $\Psi(t)$ satisfying

$$|\Phi(0) - \Psi(0)| < \delta \tag{2.54}$$

also satisfies

$$|\Phi(t) - \Psi(t)| < \epsilon, \quad t > 0. \tag{2.55}$$

---

[63] See R. J. Knops, ed., *Symposium on Non-Well-Posed Problems and Logarithmic Convexity* (New York: Springer-Verlag, 1973). As an example of a non-well-posed problem, let D be a simply connected, closed, bounded region in n-space with smooth boundary δD, and consider the following problem: $Lu = [\partial u/\partial t] + \Delta u = 0$ in $D(0, T)$; $u = 0$ on $\partial Dx[0, T]$; $u(x, 0) = f(x)$, where solutions $u(x, t)$ are C^2 in D. In general, no such solutions exist; those that do, do not depend continuously on variations in the initial conditions.

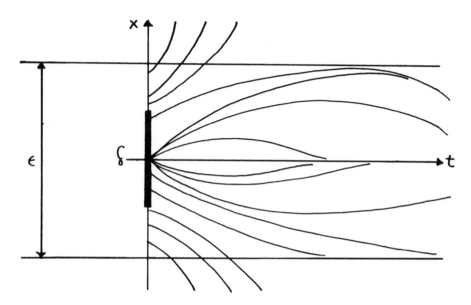

Figure 2.7
Stable solutions in XT space.

Frequently stability is specialized to equilibrium solutions,

$$f(t, 0) = 0 \quad \text{for} \quad t \geq t_0, \tag{2.56}$$

and an equilibrium solution Φ is dubbed *attractive* if there is some μ such that

$$\lim_{t \to \infty} | \Phi (t) - \Psi (t) | = 0 \tag{2.57}$$

for *all* $| \Psi (0) | < \mu$, *asymptotically stable* if it is both stable and attractive.

This is stability in the sense of Liapunov: the geometrical content behind the concept may be exhibited by depicting solutions to (2.52) as curves in an $(n + 1)$-dimensional XT space, as in Figure 2.7.[64]

Classical discussions of stability stick pretty much to second-order

[64] For a thorough introduction to the mathematics behind these methods, see W. Hahn, *Stability of Motion* (New York: Springer-Verlag, 1967). A much shorter and more readable account is available in J. La Salle and S. Lefschetz, *Stability by Liapunov's Direct Method* (New York: Academic Press, 1961); the examples that I mention here are discussed on pp. 32, 33. Hirsch and Smale offer a briskly up-to-date discussion of issues in the qualitative theory of stability in their text *Differential Equations, Dynamical Systems and Linear Algebra*.

and autonomous systems of ordinary differential equations; these are represented in the Poincaré phase space, where solutions appear as velocity vectors, equilibria as singular points. This makes for an especially attractive geometrical portrait of stability. Thus solutions to the stable system

$$\frac{dx}{dt} = y, \qquad \frac{dy}{dt} = -x, \tag{2.58}$$

describe concentric rings about a singular point (Figure 2.8a); solutions to the asymptotically stable system

$$\frac{dx}{dt} = -x, \qquad \frac{dy}{dt} = -y, \tag{2.59}$$

rays tending toward a center focus (Figure 2.8b); and solutions to the unstable system

$$\frac{dx}{dt} = x, \qquad \frac{dy}{dt} = y, \tag{2.60}$$

rays leaving a singular point (Figure 2.8c).

The concept of stability just sketched is specialized to solutions of ordinary differential equations, with a given solution seeming stable just in case neighboring solutions stick close to it as time goes on. Such solutions are actually functions, of course, and some understanding of the space in which such functions are evaluated is taken for granted. Here the appropriate space is R^n, on which a norm and inner product have been defined.

Solution-by-solution appraisals of stability, however, lead to a natural

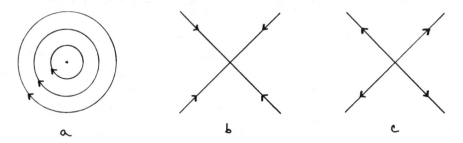

Figure 2.8
(a) A stable system. (b) An asymptotically stable system. (c) An unstable system.

generalization. Stability is a concept that in its fundamentals involves first an object and then some evaluation of the behavior of similarly situated objects, given some measure by which similarity in situation can be reckoned. Instead of taking as basic the individual solutions to a system of ordinary differential equations, Andronov and Pontryagin proposed taking the dynamical system itself (in the sense of p. 48, above) as the stable object. This is *structural stability*, the term that has come to supplant their original *systèmes grossiers*; what is involved here is not the behavior of solutions near the initital point of an arbitrary solution, but the characteristics of the entire family of solutions to a given system of equations. Informally, a dynamical system $< P, V >$ is said to be structurally stable if a slight perturbation of the system, induced perhaps by an alteration of the equation's parameters, yields a system $< P, V + \Delta h >$ that is in a topological sense close to the original system.

Andronov and Pontryagin defined structural stability in the plane; so did Peixoto. A more general definition is obtained via the assumption that a given dynamical system has been laid out over a finite- or infinite-dimensional manifold. This brings the subject of structural stability to those surfaces that are most useful in qualitative dynamics. Thus consider the particular system $< P, V >$ defined over a manifold M. The integration of the various vector fields, of course, locally defines the solutions to an ordinary differential equation; these may be elegantly understood as a set of one-parameter diffeomorphisms $\gamma_t : M \to M$ of the manifold onto itself. Such is the *flow* of the differential system.

Now the analysis of stability makes sense only relative to some measure of distance. Structural stability is a concept affirmed of dynamical systems taken as a whole; what is needed, therefore, is the set of all dynamical systems together with a metric measuring the distance between pairs of systems. The standard metric has been framed in the C^1 topology, with dynamical systems distinguished in virtue of a distance measure defined on their associated vector fields:

$$\rho[X_0, X_1] = \sup_{M} \left\{ |X_0 - X_1| + \sum \left| \frac{\partial X_0}{\partial x} - \frac{\partial X_1}{\partial x} \right| \right\} .$$

A dynamical system $X = < P, V >$, then, is structurally stable if for any $\epsilon > 0$, there is a $\delta > 0$ such that for any system $Y = < P', V' >$ with $\rho[X, Y] < \delta$, there is an associated homomorphism $\zeta : M \to M$ such that

1. ζ maps trajectories of X onto trajectories of Y;
2. ζ is an ϵ-homomorphism, i.e., for every point p on M, $d[p, \zeta(p)]$
   $< \epsilon$, where d is now the Euclidean distance between points.[65]

Structural stability is plainly a global concept: systems that are structurally stable—the van der Pol equation is an example—preserve the qualitative characteristics of their entire phase portrait in the face of perturbations. The simplest system *not* structurally stable, by way of contrast, is the harmonic oscillator (2.58). Arbitrarily small perturbations (i.e., the introduction of friction) will change the behavior of the origin from a sink to a saddle to a source; each change represents a topologically decisive violation of the system's dynamical character.

In defense of structural stability, Professor Thom has recourse to a brief if measured reflection on the conditions "de l'observation scientifique." The basic faith behind all scientific activity is expressed by the conviction that experimenters experimenting in much the same way with much the same system will come up with much the same results. All experimentation, however, conducted with whatever scrupulous objectivity, involves some minimal residue of interaction between the observer and the processes being observed. These interactive irritations correspond to perturbations inflicted on a dynamical system:

On ne peut donc espérer obtenir des résultats approximativement égaux ..., que si l'on admet implicitement que l'évolution du système S à partir de l'état α présent au moins une stabilité qualitative par rapport aux perturbations des conditions initiales et du milieu extérieur. Ainsi donc l'hypothèse de stabilité des processus scientifiques isolés apparît comme un postulat implicite de toute observation scientifique.[66]

Principles sustained by arguments of this sort have often been called *regulative*—largely by philosophers who would be embarrassed to have them called metaphysical. Nor are the arguments themselves inescapably compelling. A theory that is satisfied in models that are not structurally stable may well be scientifically inadequate, but the skeptic will ask whether nature necessarily had scientific adequacy in mind in the creation of physical systems; it makes sense to press for structural stability only if one believes that there are no interesting systems that *must* be modeled by a set of structurally unstable equations. But then structural stability would seem to be an empirical consequence of scientific investigation and not a precondition for its creation.

---

[65] See M. M. Peixoto, "On Structural Stability," *Annals of Mathematics*, Vol. 69, No. 1 (1959), pp. 189-220, for details.
[66] René Thom, *Stabilité Structurelle et Morphogènese* (New York: W. A. Benjamin, 1973), p. 33.

Metaphysical reservations aside, structural stability is still too strong a constraint. Consider the set Σ of structurally stable systems: this will be a collection of mathematical objects preserving qualitative features under perturbation. Do the stable systems form an open and dense subset in the space of all dynamical systems? If so, they are *generic*, and the restriction to stable systems may be dropped: a system that is frankly unstable may be arbitrarily approximated if it is embedded in a set of dynamical systems where the stable systems form a generic subset. Although the harmonic oscillator is unstable when its coefficient of friction $b = 0$, one can come as close as one pleases to *this* system by letting b creep to zero from above. What is really of importance is the *placement* of unstable and stable systems in the set of all dynamical systems.

For a time, it was hoped that structural stability would prove generic in every dimension. Peixoto showed that this was in fact the case for DIM $\leqslant 2$. But Smale and Williams have demonstrated that the reverse is true in higher dimensions, so for multidimensional dynamical systems there are no *general* assurances that the unstable systems can be approximated by the stable ones.[67]

If stability is in one sense too strong a requirement for physical usefulness, in another sense it is too weak. In all the concepts of stability so far considered, the essential idea has been to investigate the mathematical behavior of objects "close" to a given object. If the behavior is similar, on some established measure of similarity, there is stability. But what counts as an object *near* a given object is not absolutely fixed. Instead, nearness is *functional*: the solutions near the initial point of an arbitrary solution will, in general, be determined by the overall distance between solutions that one wishes to preserve.

There are cases in which a system performs in what is intuitively a stable fashion even though its behavior cannot be brought arbitrarily close to fixed limits. Aircraft, for example, frequently yaw in a perfectly stable way about a particular region of disturbance. And there are cases in which a system is stable only within a region that is inaccessible to measurement. The dependence of the definition of asymptotic stability on some particular choice of μ might make the point vivid: take μ small enough and the system that results may for all intents and purposes exhibit an almost intolerable degree of instability—this despite the fact that wedged close to the origin, solutions may start to drift inward toward an equilibrium that is asymptotically stable.

---

[67] S. Smale, "Structurally Stable Systems Are Not Dense," *American Journal of Mathematics*, Vol. 86 (1966), pp. 491–496.

La Salle and Lefschetz, for example, look to a notion of stability that will count as stable processes that return tolerably close to a point of equilibrium without actually approaching that point continuously.[68] They have oscillating systems chiefly in mind.

In detail, their definition is this. Consider a simple system

$$\frac{dx}{dt} = f(t, x), \quad t \geq 0, \tag{2.61}$$

that has an equilibrium point at the origin for all $t \geq 0$. Consider next the system that results when (2.61) is perturbed:

$$\frac{dx}{dt} = f(t, x) + p(t, x), \quad t \geq 0. \tag{2.62}$$

Here $p(t, x)$ denotes the alteration of the original equation: keeping in mind Professor Thom's strictures on the inevitability of error in any method of measurement, we can take (2.62) to be the system one hits on in those inevitably careless attempts to get (2.61) itself. Three items must now be fixed: a closed and bounded set Q containing the origin; a subset Q_0 of Q; and a bound δ on the set P of acceptable perturbations. For each $p \in P$,

$$|p(x, t)| \leq \delta \text{ for all } t \geq 0 \text{ and all } x. \tag{2.63}$$

Now if for every $p \in P$ and all $t_0 \geq 0$, each solution $\Phi(t)$ of (2.62) with initial condition $\Phi(t_0) = x_0$ remains in Q for all $t \geq 0$ whenever x_0 is in Q_0, the origin at (2.61) is said to be *practically stable*. This is the situation represented by Figure 2.9 and illustrated somewhat more intuitively in a neat diagram drawn by W. Hahn (Figure 2.10): the ball in the valley, while stable, is practically unstable; the ball on the hill, while unstable, is practically stable.

Meteorology is a science much involved with concepts such as practical stability; it is a science, too, that shares some problems with the social sciences. As a practical matter, meteorologists must derive their predictions from a fairly sophisticated theory dealing strictly with the hydrodynamics of the atmosphere. Standard forecasting techniques subdivide the region for which a forecast is being prepared. Various points on the grid correspond to data stations, and the initial state of a meteorological system is built up as an aggregate of states reported at

[68] La Salle and Lefschetz, *Stability by Liapunov's Direct Method*, p. 121.

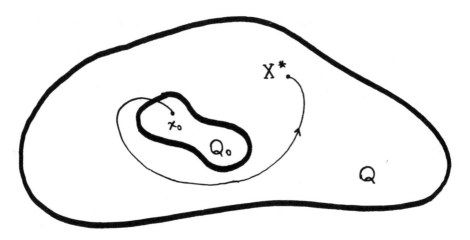

Figure 2.9
Practical stability.

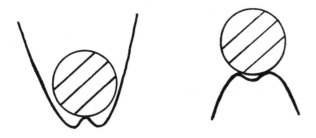

Figure 2.10
A practically stable and a practically unstable system. After W. Hahn, *Stability of Motion* (New York: Springer-Verlag, 1967), p. 8.

different points. No system of observation can report everything, and how much gets observed accurately will obviously be a function of the distance between points of observation. At any reasonable distances some processes will be dynamically inaccessible. The coarseness of the grid makes it impossible to fix the initial conditions of turbulences that take place on a scale vastly smaller than the scale used in setting up the forecasting grid. If two weather stations are five miles apart, neither will report the eddy of wind or the gusting of smoke at a point precisely halfway between them. This is a physical and not a mathematical limitation:

If we could precisely fix the initial state of all the small-scale motions and obtain the exact solutions for their dynamical equations, then, to put it abstractly, the periods of predictability would be in no way limited. . . .

But we can fix the initial values of the meteorological fields only at points of a much coarser grid . . . so that the individual motions with scales smaller than the interval are obviously not fixed at all; in addition, we make random errors in measuring, interpolating, and smoothing. Because of these initial errors, even with an exact solution of the exact dynamic equations, a forecast inevitably contains errors that, generally speaking, will be larger the longer the period of the forecast.[69]

Having a dynamical model, the meteorologist has a mathematical object that represents *at best* a perturbed or displaced condition of the true weather system. That such a model begins to deviate from the system it is meant to predict need not reflect any intrinsic instability in the atmosphere: the smallest distances sanctioned by the model may well be gross enough to send solutions smashing through reasonable bounds in a very short time. At accessible distances such stable systems may well be practically unstable.

Nor need the errors be all that gross. Kolmogorov imagined two planets with completely identical weather systems. On the first a handkerchief flutters. How long, he asked, would it take for the weather on these planets to become completely different? That the weather does become different is evidence for instability; that there is some fluttering too small to affect the weather at all is evidence that the instability is practical. The fact that there are some distances within which such systems stay stable hardly means that there is anything like a reasonable hope for framing models that stick to stability-preserving distances. There is very little difference between being impossible and being merely difficult so long was what is difficult is difficult enough.

This general class of problems Ralph Abraham assembles under the title *the problem of error.*[70] Typical cases occur simply because the dynamical system perfectly matched to reality is for one reason or another intellectually inaccessible. In economics, for example, there is the problem of aggregation: to get at the true state of the economy requires a dynamical model taking into account the economic activities of each consumer. Obviously this is impossible, so consumer demands are aggregated. Save for the simplest cases, this leads inevitably to biases. Whether such distortions in the econometrician's working model generate a system of errors that systematically separates the model from reality is a function of the size of the aggregation errors and the underlying stability of the working model. The problem of aggregation is simply a theoretical image of the meteorologist's problem in selecting

[69] Monin, *Weather Forecasting as a Problem in Physics*, pp. 139-140.
[70] Abraham, *Foundations of Mechanics*, p. 231.

a suitable scheme to cover with observation stations some vast and physically remote portion of the earth's surface.

The definition of practical stability offers a purely qualitative understanding of stability taken in a way that makes all the obvious concessions; still, the definition is couched in terms of three parameters —the first fixing the acceptable perturbations, the second the acceptable distances between initial states, and the third, and most important, the set of acceptable states themselves—and nothing in the notion of stability itself suggests how these might be defined.

The region bounded by Q, for example, might be understood as constituting the set of predictions better than some arbitrary prediction. Q itself can be taken as the farthest point at which a set of predictions retains anything like qualitative usefulness. For frankly unstable processes, such as those found in meteorology, it becomes important first to make sense of the notion of an acceptable state, and second to have some indication of the time it will take an arbitrary and unstable solution to meander out of the zone of acceptability.

One rough and ready way to measure the limits of predictability is to compare dynamical processes predicted by a given model with the estimates that might be yielded simply by guessing. This is the problem as set out by P. D. Thompson in a beautiful and perspicuous paper:

Let us suppose that the general hydrodynamical equations are integrated . . . starting with two *different* initial states. In one case, we shall begin with the true state and in the other, with the reconstructed state, consisting of the true state plus a field of analysis error that is random with respect to the true state. In the first case, the prediction must be correct by hypothesis. In the second, the prediction is *incorrect initially*, and will generally differ from the correct one at all later times. The amount by which it differs at various times after some specified initial instant is, of course, a measure of unpredictability or error. . . . Thus the difference between two solutions of the hydrodynamical equations, starting with two initial states that differ by a random field error, is a measure of the *essential* unpredictability of the atmosphere. In particular, if the average magnitude of such differences at some later time approaches the average error of guessing, the atmosphere has become essentially unpredictable beyond that time.[71]

The two equations that Thompson mentions correspond to (2.61) and (2.62) in the definition of practical stability; when the difference between them approaches the average error of guessing, solutions at hand will have passed across the boundary of acceptable states marked

[71] P. D. Thompson, "Uncertainty of Initial State as a Factor in the Predictability of Large Scope Atmospheric Flow Patterns," *Tellus*, Vol. 9 (1957), p. 3.

by Q in the same definition. Thompson goes on to derive an analytic estimate for the degree of atmospheric unpredictability using Charney's modified model of the Navier-Stokes equations.

For systems of some instability, a reckoning of what might be dubbed the *average escape time* will no doubt prove important: this might be a probability density defined over the entire function space of solutions. Should the average escape time prove rather short, forecasting might be a simple impossibility since errors might grow at a speed faster than the speed at which computers can integrate the appropriate dynamical equations.

Quite another, but obviously related problem arises when a given model is *embedded* in the context of a larger model. Consider the set of physical observations to be points in a given topological space D; associated with each such D is a specific mathematical model M_d taken together with a prediction P_d, where $d \in D$ represents a fixed point in the topological space. The problem of embeddability arises, Abraham remarks, because

> we may never account for the full complexity of the experimental domain in the family of mathematical models $F = \{d, M_d, P_d\}$. The choice of observables represented by points $d \in D$ excludes many that might be incorporated. If we enlarge the family F to another $F' = \{d \in D', M_d, P_d\}$, where D is a subset of D', then a particular model M_{d_0} might be stable relative to F—but not stable relative to F'. In this case, the prediction is approximately preserved by some variations in the observable parameters and wildly disturbed by other variations.[72]

An obvious and simple case, again, is the harmonic oscillator, which is stable relative to a zero coefficient of friction and *any* small change of mass, but profoundly unstable once the coefficient of friction changes slightly. Under these conditions, original vouchers of stability must be augmented by guarantees that stability does not disappear under extension.

What I have called the problem of embeddability, Ando and Fisher call the problem of partition: theirs is explicitly a concern for "the conditions under which partial dynamics—or indeed the dynamics of the economic system as a whole—can be treated without explicit concern for the larger system of socio-political equations in which a particular system is embedded." What they mean by partition is the degree to which untoward causal influences running from the larger system to the smaller system may be safely ignored. The most interesting theorem

[72] Abraham, *Foundations of Mechanics.*

in this area is Ando and Simon's result, which says roughly what one would expect—that if the dynamical influences on a partitioned system are weak, the system will behave almost as if it were uninfluenced.[73] The theorem is somewhat less interesting than it might be because it holds only for systems with decomposable matrices; this is a restriction pretty much on all fours with the curtailment of the observation that all men are mortal strictly to albinos.

In place, then, of a simple line severing the physically useful models from all the rest, there is a contrapuntally organized group of considerations, with structural stability a leading theme, but modified now by reservations that pertain to the existence of generically stable systems, that treat the problem of practical instability, and that recall restrictions on embeddability.

The question to which I have been tending is whether mathematical models of the world, of the sort offered in *World Dynamics* and *The Limits to Growth*, are liable to be expressed by well-posed systems of ordinary differential equations.

In writing about René Thom's catastrophe theory, E. C. Zeeman argued that the social and biological sciences are marked by properties that set them mathematically apart from the physical sciences. He writes especially of the distinction between catastrophe theory and the classical, conservative theories of physics involving Hamiltonian mechanics. The properties that mark Thom's theory and that are notably absent from classical physics are *discontinuity* and *divergence*. By discontinuity, Zeeman means that in catastrophe theory a slow passage through a space of causes is liable to produce an abrupt change in a corresponding space of effects. But in a more general sense, Zeeman also sees the discontinuous as a mark of political and economic life, where gradual changes entail dramatic effects, as when nations relentlessly and imperceptibly pass over some irrevocable point of bifurcation and find themselves at each other's throats. A theory is divergent, in contrast, if small changes in its initial conditions are not preserved, with subsequent states spreading apart like a perpetually expandable accordion. Divergence, too, is a feature of Thom's theory, and one than Zeeman sees reflected in the plain instabilities of social and economic life. "Throughout the biological and social sciences," he writes, "these two qualities are frequently found, and are the origin of the label 'the inexact sciences' because it was thought that the existence of such qualities proved the

[73] H. A. Simon and A. Ando, "Aggregation of Variables in Dynamic Systems," *Econometrica*, Vol. 29 (April 1961), pp. 111-138.

impossibility of finding mathematical models that could be used for prediction."[74]

Thus, from Zeeman's perspective, systems of ordinary differential equations that are well-posed in anything like my augmented sense cannot faithfully reflect the intrinsic discontinuity and instability of social and political life. And this is an objection in principle to the mathematical methods that make up the models of the world.

[74] E. C. Zeeman, "Applications of Catastrophe Theory," *Bulletin of the London Mathematical Society* (forthcoming).

3 Mathematical Systems Theory

The terms "systems," "systems concepts," "systems approach," and "systems science" are used so widely and so broadly today that they tend to connote fuzzy thinking.

R. E. Kalman, P. L. Falb, and M. A. Arbib[1]

Introduction

Mathematical systems theory, like systems theory generally, is an object of uncertain purity. Such journals as *Mathematical Systems Theory* give the reader no secure sense of their subject: articles on automata abound, but there are also pieces on the theory of ordinary differential equations, dynamical systems, filters, cascaded transducers, dynamical polysystems, cellular automata, automatic control, maximum control theory, stochastic control, linear systems theory, regeneration theory, probabilistic automata, stability theory, state-space analysis, phase-space analysis, computational algorithms, adaptive control theory, Bode diagrams, Nyquist diagrams, the algebraic theory of linear systems, k-module theory, Tolerance automata, the Kalman-Bucy filter, Wiener's theory of prediction and the Wiener filter, Ho's algorithm, cascade decomposition theory, and the theory of input-output systems.[2]

Not to mention reliable automata, self-reproducing automata, optimal and self-optimizing control theory, the theory of nonlinear control, biological control theory, minimization theory, dynamic programming, the bang-bang control problem, differential games, the theory of nonlinear oscillations, the time-optimal control problem, the calculus of variations, Pontryagin's maximum principle, Hamilton-Jacobi theory, linear systems and the quadratic criterion, root-locus analysis, and Nichols's chart design.

Mathematical systems theory arose as abstract engineering. The years between 1915 and 1950 or so were decisive in the expansion of the engineering disciplines. Almost all of the great work was done within electrical and acoustic engineering. Richard Bellman's collection lays the fundamental papers out seriatim: Minorsky on self-excited oscillations; Wiener on the filters; van der Pol on forced oscillations. Since then the subjects have grown considerably. From one perspective growth has been strictly linear: classical linear theory → stochastic control theory under the influence of Wiener and Kolmogorov → the theory of linear prediction → the development of optimal control theory and the Pontryagin principle → nonlinear control theory → and, looking

---

[1] R. E. Kalman, P. L. Falb, and M. A. Arbib, *Topics in Mathematical System Theory* (New York, McGraw-Hill, 1969), p. 3.

[2] For a recent collection, see Richard Bellman, ed., *Mathematical Trends in Control Theory* (New York: Dover, 1964). There is also the important journal, *Mathematical Systems Theory*, published regularly by Springer-Verlag, and the *SIAM Journal of Control*. For a personal memoir covering the rise of modern control theory, see Harman Parish, "Those Fabulous Filters: Four Decades with Oswango Labs," Oswango Technical Report No. 67–3 (Oswango Trade Publications, 1967).

now to the future, nonlinear stochastic control theory and adaptive control theory. This suggests a straight line extending to infinity, but one of the interesting developments of the last decade has been the reintroduction of classical results by a culling of generalizable elements from concrete cases; and this calls the closed circle to mind. Thus, as standard techniques were extended first from strictly linear to stochastic control and then to systems of adaptive control, the old-fashioned linear case itself was redescribed in modern algebraic terms. Much of mathematical systems theory therefore mirrors the rise of abstract algebra. This has had some effect on the *style* of mathematics within the systems sciences, as even the most prosaic of engineering accomplishments—the design of a truss, say—has come to be expressed in the language of modules, linear transformations, and matrix algebra.

Models drawn from the engineering disciplines have always had a certain fascination for the social scientist, though this admiration has often been directed toward techniques that the mathematician looks upon with paternal embarrassment. Here I am thinking in particular of the practice in mathematical economics of reinterpreting classical multiplier theory in terms of such routine items as the transfer function and the like.[3] But, plainly, the engineering sciences throw a general *philosophical* attraction toward the social scientist: dealing with a realm of objects that at least in part evince something like human powers and capacities—machines do work, automata compute—they hold out the eternal promise of a system of thought sturdy enough to encompass both human agents and the curious mechanical constructions that they fashion in their own image.

Mathematical systems theory depicts a system as something roughly three-staged, with an input and an output flanking a conversional device of varying capacities—a filter, perhaps, as in Figure 3.1, where both input and output have been depicted as functions of time. What stands between them is opaque—a quintessential black box. Obliterated are distinctions between electrical, mechanical, thermal, or social systems. If the system's inscrutability is scrupulously respected, the engineer is left with only inputs and outputs; the analysis of what results has to be based very narrowly on a record of what happens to the device or plant and what the plant does when what happens happens.

---

[3] See R. G. D. Allen, *Mathematical Economics* (New York: St. Martin's Press, 1959), or Oskar Lange, *Introduction to Economic Cybernetics* (New York: Pergamon Press, 1970). For algebraic techniques, see R. E. Kalman, "Algebraic Aspects of the Theory of Dynamical Systems," in J. K. Hale and J. P. La Salle, eds., *International Symposium on Differential Equations and Dynamical Systems* (New York: Academic Press, 1967), pp. 133–146.

Figure 3.1
The systems analyst's system.

The resulting analysis is called, naturally enough, an *input-output description*, and the appropriate mathematical model is usually an integral equation.[4]

In the case of classical control theory, inscrutability is compromised: assumed is some explicit governance of the black box by a system of differential equations. A complete specification of the *dynamical* properties of the system of Figure 3.1 might be expressed by

$$\frac{dy(t)}{dt} = A(t)y(t) + B(t)x(t), \tag{3.1}$$

which acts to glue $y(t)$ to $x(t)$. The fundamental systems here are much like *dynamical systems*. The method by which engineering objects are classified, moreover, mirrors the organization of the theory of differential equations. But it would be a mistake to dismiss the differences between engineering and dynamical systems: engineering is preeminently a concern for the relationship between inputs and outputs, while the mathematical focus of the theory of dynamical systems is on the equations themselves and the models in which they are satisfied. Equation (3.1) constitutes a dynamical system; (3.1) enriched by equations expressing the relationship of changes in output to changes in the system's dynamical properties becomes part of mathematical systems theory.

Mathematical systems theory splits in two, with the concept of continuity marking a natural line of division. Mathematical systems that

[4] See Jan C. Willems, *The Analysis of Feedback Systems* (Cambridge, Mass.: The MIT Press, 1971). The great figure here is Nyquist; Willems credits the modern development of input-output theories to Sandberg and Zames. See, for example, G. Zames, "On the Input-Output Stability of Time-Varying Non-Linear Feedback Systems," Parts I and II, *IEEE Transactions on Automatic Control*, Vol. AC–11 (1966), pp. 228–238, 465–476.

embody something like a classical dynamical system and that treat time as continuous fall to one side; those that consider sequences of steps taking place at fixed intervals—as in automata, for example—fall to the other. An ordinary computer program is an example of the latter: inputs are items punched on a card; outputs are computations; and connecting them are algorithms. The computer itself is clearly an automaton, and automata are input-output devices, to put the matter crudely, whose transition tables move inputs to outputs in a discrete shuffle. A machine built along such lines fits Figure 3.1, but subjects devoted to such devices—automata theory, the theory of recursive functions—do not travel with the old familiar engineering crowds. Nevertheless, automata theory and recursive function theory have been subsumed by the journals, and systems theorists now accept definitions of systemhood broad enough to encompass any device taking inputs to outputs by means of a set of states.[5] The theorems that have resulted are, to be sure, meager enough; but what counts, the systems theorists argue, is the high degree of conceptual coherence that the modern methods bring.

[5] See, for example, M. A. Arbib, "Automata Theory and Control Theory: A Rapprochement," *Automatica*, Vol. 3 (1966), pp. 161–189.

Linear Systems, Classical Control Theory, and Political Science

There are, generally speaking, two circumstances in which it is difficult to analyse mathematically a social system: the first is when the system is not linear; the second is when it is.

B. L. Hendricks[6]

The engineer aims to understand the plant as an input-output device. Knowing what goes in, he wishes to know what comes out. Or vice versa. Man, Richard Bellman has remarked, studies science for reasons of both understanding and control. This is forthright, if a trifle vulgar. But the stress on the usable usufructs of his daily toil serves to distinguish the engineer from the physicist or applied mathematician whose work he often appropriates. To the theoretician goes the interesting business of figuring out the dynamical laws that animate a plant; to the engineer, the task of putting those laws to good use in matters pertaining to control, coordination, regulation, and design. These are the quintessential engineering concepts: seeing them thus spread out points to another distinction to which mathematicians habitually repair, that between pure and applied mathematics. It is a distinction, however, without a difference: systems, too, are theoretical objects, although of a kind not generally found in pure mathematics. Confronted with blueprints and the request that points A and B be spanned by a suspension bridge, the engineer whose efforts have extended mainly to perturbation techniques in Banach spaces will react with all the blank incompetence of the mathematician trained strictly in algebraic k-theory.

That there is nothing invidious in the kinds of objects regularly falling under engineering scrutiny hardly means that the engineer and the pure mathematician study the plant in ways that are in some sense parallel, along lines suggested by, say, the microbiologist and the molecular biologist. The mathematician sees the plant as a dynamical system; the engineer, as an input-output device. *The first concern is logically prior to the second.* It is generally reckoned no part of the engineer's professional obligations to get at the physical, chemical, social, political, or biological laws that describe dynamics. This usually gets a standard yawn of assent from social scientists, but then why should engineering concepts prove so implacably attractive to them when the great problems in the mathematical analysis of the social sciences all have to do with suitably uncovering their dynamical structure?

[6] B. L. Hendricks, "Some Confusions in the Qualitative Analysis of the Social Sciences," in B. S. Latvoks, ed., *Temporal Measures of Uniform Efficiency: Essays in Honor of Roy Blatkin* (San Francisco: The Maypo Press, 1967), pp. 312–313.

Abstract engineering is conceptually dominated by *analysis* and *design*, which should not be surprising since engineers, after all, are expected to create a class of objects that sooner or later perform in some specified manner. And much that is of interest in the current engineering literature, such as *control theory*, falls legitimately within the province of the abstract theory of design.

Control is a concept of four parts: first, there is the system, the object to be controlled; second, the objectives of the system, which are what the system is controlled for; third, the controls themselves, which are what the system is controlled with; and, finally, the measure of relative optimality, which is what the controls cost. A *control system* is a device of such parts; picture it a system, in the sense of Figure 3.1, to which a *regulator* has been attached, where a regulator is a mechanism controlling the system at fixed costs. This sounds like a prospectus for *optimal control theory*, and it is. Classical control theory is optimal control theory truncated to a subject that in its modern formulation has to do largely with the *existence* of suitable controls, without regard to costs.

Like so many other distinctions, the division between analysis and design disappears on the theoretical level, although it surely makes sense to talk about analysis *or* design on a case-by-case basis. Thus one could talk about linear control theory as a brilliantly abstract prolegomenon to the theory of design: in determining just *how* the movement of a rocket might be controlled—a problem of design, clearly—it is obviously helpful to know first whether it may be controlled at all—a problem in classical control theory. Equally, I should think, both classical control theory and optimal control theory might be envisaged as pure mathematical subjects surveying abstract objects about which interesting *existence* theorems are proved.

I bring this point up only to narrow appropriately the range of my own interests in examining the suitability of engineering models for the social and biological sciences. Certain social systems arise as the result of conscious design—the Federal Reserve Board, or the hierarchy of a corporation. For such systems, theoretical results in optimal control theory may have some practical implications. Given a certain range of economic choices, optimal control theory offers advice. With this I have no quarrel. Such questions of design, however, are quite distinct from the question of whether social or political or biological systems actually satisfy the assumptions of optimal control theory. *That* is not at all a matter of extracting from an engineering discipline some suitable principle for design; rather, it is a basic, and unresolved, question of *analysis* itself.

The distinction may be set out more clearly if we look to the general schema of optimal control. What is given, abstractly, is the dynamics of the plant,

$$\frac{dx_i}{dt} = f_i(t, x_1, x_2, \ldots, x_n, u(t)), \qquad i = 1, 2, \ldots, n, \tag{3.2}$$

where $u(t)$ is a control variable and the x_i are the state variables of the plant. Optimal control theory involves the search for a control function that will optimize a criterion or cost functional, usually defined as an integral quadratic function of the x_i and u. Pontryagin obtained a complete solution to this problem.

That $y^*(t)$ happens to comprise an optimal trajectory for a given social system is usually a matter of design: in the course of things, $y^*(t)$ is most often exhibited because some or another planner proposes a specific solution to a specific problem—proposes, in short, to design, at least partially, a given system in a given way. But that the description just given embodies the lineaments of an abstract object suitable for the modeling of a social system is not a matter of design generally: the social system is given, or proposed; matching it to the assumptions of optimal control theory is essentially a problem of analysis.

Transform and Transfer

The classical techniques in the engineering sciences display a purity of achievement made possible only by the most severe simplifications. The plant is restricted to systems governed by autonomous linear differential equations with constant coefficients. Thus, in considering possible dynamical systems for instantiation at Figure 3.1, there is a downward progression, like a descending scale, from the general case

$$\frac{dx_i}{dt} = f_i(t, x_1, x_2, \ldots, x_n, u(t)), \qquad i = 1, 2, \ldots, N \tag{3.3}$$

to the linear case

$$\frac{dx_i}{dt} = \sum_{j=1}^{N} a_{ij}(t)x_j + b_i u(t), \qquad i = 1, 2, \ldots, N \tag{3.4}$$

to the autonomous case

$$\frac{dx_i}{dt} = \sum_{j=1}^{N} a_{ij}x_j + b_iu(t), \qquad i = 1, 2, \ldots, N \tag{3.5}$$

to the single-input case

$$\frac{dx}{dt} = ax + bu(t). \tag{3.6}$$

That (3.6) rather than (3.3) expresses the dynamical properties of Figure 3.1 makes possible the most notable of simplifications: here the classical engineering tradition uniformly commends the *Laplace transform*; this is essentially a way of replacing analytic by algebraic techniques and then expressing what results in terms of *transfer functions*.

Laplace transforms[7]
If $f(t)$ is a real-valued function defined for $t > 0$, the Laplace transform $L[f(t)] = F(s)$ is defined by the relationship

$$L[f(t)] = F(s) = \int_0^\infty f(t)e^{-st}dt, \tag{3.7}$$

where s is a complex variable. The Laplace transform passes the mathematician from the real to the complex plane and thereby shifts analysis to the *frequency domain*. Having reset a given equation in the complex plane, one must, of course, eventually convert back to the real or *time domain*; the right tool is the inverse Laplace transform,

$$L^{-1}[F(s)] = f(t) = \frac{1}{2\pi i} \oint F(s)e^{st}ds, \tag{3.8}$$

where $\oint$ is a *contour* or *convolution* integral. In practice, it is rarely necessary to use such heavy machinery: computational techniques are available instead.

[7] A good introduction to the Laplace transform is DiStefano, Stubberud, and Williams, *Feedback and Control Systems* (New York: McGraw-Hill, 1967), a Schaum's outline, no less. For classical techniques, see Richard C. Dorf, *Modern Control Systems* (Reading, Mass.: Addison-Wesley, 1967).

Consider, for example, an nth-order linear differential equation with constant coefficients,

$$\sum_{i=0}^{n} a_i \frac{d^i y}{dt^i} = x(t), \tag{3.9}$$

with initial conditions that at $t = 0$

$$\frac{d^k y}{dt^k} = y_0^k, \qquad k = 0, 1, \ldots, n - 1. \tag{3.10}$$

To obtain a formal solution, transform (3.9), which becomes

$$\sum_{i=0}^{n} \left[a_i \left(s^i Y(s) - \sum_{k=0}^{i-1} s^{i-1-k} y_0^k \right) \right] = X(s) \tag{3.11}$$

or

$$Y(s) = \frac{X(s)}{\displaystyle\sum_{i=0}^{n} a_i s^i} + \frac{\displaystyle\sum_{i=0}^{n} \sum_{k=0}^{i-1} a_i s^{i-1-k} y_0^k}{\displaystyle\sum_{i=0}^{n} a_i s^i}. \tag{3.12}$$

The solution can now be expressed in terms of the inverse Laplace transform:

$$y(t) = L^{-1} \left[\frac{X(s)}{\displaystyle\sum_{i=0}^{n} a_i s^i} \right] + L^{-1} \left[\frac{\displaystyle\sum_{i=0}^{n} \sum_{k=0}^{i-1} a_i s^{i-1-k} y_0^k}{\displaystyle\sum_{i=0}^{n} a_i s^i} \right]. \tag{3.13}$$

The matched terms between brackets are known as the *forced* and *free* responses of the system, respectively: the first is the solution when $y_0 = 0$; the second when the input function $x(t) = 0$.

Transfer functions

Although transform methods are useful when a solution in the time domain is required, the analysis of an electrical, thermal, or mechanical system generally stops with the forward Laplace transform, with the various transformed variables united in an overall *transfer function.* Most simply put, the transfer function for a system such as that defined by (3.9) is the *ratio* of the Laplace transforms of input and output variables:

$$G(s) \equiv \frac{Y(s)}{X(s)}. \tag{3.14}$$

This leads to an easy representation of a system-as-a-whole in terms of *block diagrams,* such as Figure 3.2 for (3.9) or Figure 3.3 for some hypothetical system with two inputs and two outputs. All variables, of course, are complex; symbols within blocks represent transfer functions, and the overall effect of the block diagrams may be conveyed by equations. Thus Figure 3.3 is equivalent to

$$Y_1(s) = G_{11}(s)X_1(s) + G_{12}(s)X_2(s),$$
$$Y_2(s) = G_{21}(s)X_1(s) + G_{22}(s)X_2(s). \tag{3.15}$$

Feedback may be brought within the orbit of the block diagram as in Figure 3.4, which represents the *canonical closed-loop, negative-feedback system.* Here the output $C(s)$ has been connected, via a transfer function $H(s)$, to the original input $R(s)$; the line going from $C(s)$ through the transfer function $H(s)$ onto the variable output $B(s)$ may be thought of as a kind of shadow system dogging the original open loop going from $R(s)$ to $C(s)$. It is this shadow system that gives rise to the error signal

$$E(s) \equiv R(s) - B(s)$$
$$= R(s) - H(s)C(s). \tag{3.16}$$

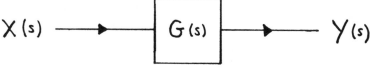

Figure 3.2
A simple system with the plant represented by a transfer function.

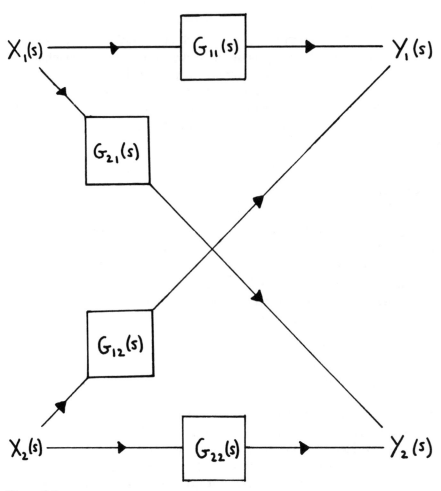

Figure 3.3
Block diagram of a system with two inputs and two outputs.

As with Figure 3.2, the effect of the block diagram of Figure 3.4 may be expressed in equational form; the output of the system is hooked directly to the error signal by a transfer function

$$C(s) = G(s)E(s) \tag{3.17}$$

or, using (3.16),

$$C(s) = G(s)[R(s) - H(s)C(s)]. \tag{3.18}$$

The overall transfer function, then, is

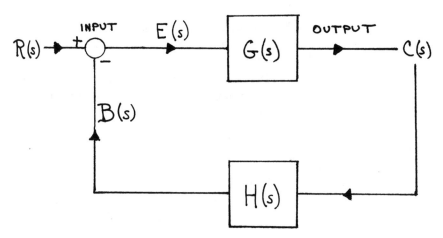

Figure 3.4
The canonical closed-loop feedback system.

$$\frac{C(s)}{R(s)} = \frac{G(s)}{1 + G(s)H(s)},$$

(3.19)

and it is this expression that sums up the chief dynamical properties of a closed-loop negative-feedback system of the sort described in Figure 3.4.[8]

In what follows, then, I take the feedback loop of Figure 3.4 as the basic object of the classical tradition in engineering, i.e., a canonical closed-loop feedback system, expressed in the notation and terminology of transfer functions and block diagrams and made up ultimately of elements whose dynamical constituents satisfy an appropriately restricted ordinary differential equation. The assumption throughout is that Laplace transforms have been used to obtain transfer functions.

Control Theory

In Figure 3.4 the theory of linear systems passes imperceptibly over to the theory of control and regulation. Actually, the relationship between the two disciplines is that of whole to part: moving from Figure 3.2 to Figure 3.4 represents really a *contraction* in scope and a specialization in content.

Feedback control offers gains but also disadvantages. There is, of course, the overwhelming advantage, once feedback has been intro-

---

[8] See Dorf, *Modern Control Systems*, pp. 41–45, for further details. Equation (3.19) is often referred to as the *fundamental formula of control theory*.

duced, of having a device that tracks a specified input function on the basis of output alone; this gives rise to the possibilities of purely automatic control, as in missiles, self-tuning amplifiers, and the like. Then there are minor measures: feedback systems are less vulnerable to variations in parameters; they are less vulnerable, too, to noise and a variety of unwanted disturbances; and feedback, finally, makes desirable transient and steady states more available than they might otherwise be. Against this achievement in control, there is a certain possibility for radical instability.

But with the change from Figure 3.2 to Figure 3.4 comes nothing like a unified theory: the classical tradition offers techniques cascaded one on top of the other, and gaps in mathematical rigor filled frequently with appeals to physical intuition.

Three questions figure prominently in the classical theory:[9]

1. What are the measures of system performance by which control-theoretic objects may be appraised?
2. What is the correct analysis of feedback control systems in terms of these performance measures?
3. How can the system be adjusted if its performance turns out to be unsatisfactory?

It is the third question that generally turns control theory toward *design*; the first and second involve analysis and form aspects of a single question. Overall, then, classical control theory is pretty much taken up with the following very general questions: Given a feedback system in various stages of specification, and given also some sense of the system's desired performance, *what is the system liable to do*? Doing what it does, *how can the system's performance be changed*?

Feedback control systems are often appraised in a space of four dimensions. The separate axes represent stability, steady-state accuracy, transient response, and frequency response. A system that is stable will not grow explosively as a result of variations in *bounded* inputs. Inputs are infinitely various, so some selection is usually necesssary for *test* purposes. Common choices are step functions, ramp functions, and sinusoidal functions. Given an input function, stability may be assessed by techniques such as Routh's criterion, the graphical root-locus method, Bode plots, or Nyquist stability diagrams. The distinction between a system's steady states and its transient response is the engineer's mirror of the mathematician's distinction between equilibrium and nonequilibrium solutions to an ordinary differential equation. Actually,

[9] See William L. Brogan, *Modern Control Theory* (New York: Quantum Publishers, 1974), pp. 22–42, for details.

steady states are periodic solutions: as the result of forcing terms in the original equation, asymptotic stability of the reduced equation is passed on to the periodic solutions of the complete equation. Steady-state accuracy reflects the requirement that the system's error signals $E(t)$ shrink as time goes on. Suitability of transient response, by way of contrast, reflects a true stability consideration: systems need guarantees against excessive and sudden overshoot, wild oscillations, leisurely settling times, and the like. Much the same concerns apply to suitable or satisfactory frequency responses: here the principal objects are yielding bandwidths, limited magnifications of input to output change, and so on. The language for obtaining specifications of performance in these last two areas is one of *gain* and *phase margin, delay times, bandwidth, cutoff times, resonance peaks,* and *frequency peaks.*

The analysis of systems is carried out for purposes of control; the classical tradition is in the service of *design by analysis*: for a system operating in an unsuitable manner, techniques of compensation are introduced to beef up performance. Thus two separate, if classical, techniques for control involve the accretion of additional feedback loops, or the modification of existing loops, and the addition of compensating networks to change phase characteristics in a given range. These compensation techniques are derived naturally from Nyquist-criterion and Bode-plot analyses.

I mention these ugly points only to make vivid the character of the classical tradition. It involves a mathematical style of stern limitations, and there is an ornithological variety to the techniques that get brought into the actual play of computation. The pure mathematician will see in this a deliberate cultivation of the engineer's epistemological maxim, "I don't know anything about mathematics, but I know what works." Still, it is only by tracing through the applications that one can achieve some feeling for the character of the objects that are analyzed in the classical tradition.

Most disciplines have a natural focus: the classical tradition takes as its own objects such as amplifiers, filters, acoustical relays, transducers, electrical circuits, and condensers. It may well be that other systems that satisfy Figure 3.1 satisfy Figure 3.4 too, but there is also the strong chance that they do not. It is always good policy to avoid conceptual attractions based on the most general features of a system of thought, if only because such general formulations are often without content.

To a certain extent, this is the problem one encounters in mathematical systems theory when the definition of a system is so broad, as in Figure 3.1, that dynamical systems in the sense of Poincaré and Birkhoff, finite-state automata, Turing machines, transducers, grammars, and electrical conductors are all covered. No theorems result from a definition so grandly conceived. Noticing the structural similarities between integers and polynomials, the mathematican concludes that both are rings. But the concept of a ring has independent content; there are axioms about rings, and many interesting and useful theorems follow from them. *There are no comparable axioms or theorems for systems in general.* Systems turn out to have little more in common than their systemhood.

This is well understood by sophisticated authors: having introduced systems in the fashion of Figure 3.1, Professor X will waste no time in specializing the issues, first by means of a suitable winnowing in the class of dynamical equations, and then by the use of specialized techniques that restrict operations even more thoroughly. The theorems that come into play, or the mathematical truths tacitly relied upon, specialize in a loose but unmistakable fashion the models in which the method is meant to work. Writers outside the engineering arts, captivated by Figure 3.1 and the thought that political systems are also systems that take inputs to outputs, should stick around to Figure 3.4 and see whether the activities of Congress lend themselves neatly to redescription in terms of constants of error, phase lags, bandwidths, and transfer functions.

David Easton and the System of Politics

It is David Easton's conviction that the political process embodies the political system.[10] A second edition of his influential *The Political System* appeared in 1971, and with it a rare and measured response to

---

[10] For an introduction to Easton's work, see Roger Scott, "Systems Analysis Without Tears: Easton and Almond," *Politics*, Vol. 7, No. 1 (1972), pp. 74-81; there is also Oran R. Young, *Systems of Political Science* (Englewood Cliffs, N.J.: Prentice-Hall, 1968). Professor Easton is by no means alone in espousing system-theoretic concepts in the analysis of political systems; but he was there first and has remained doggedly in place ever since. Other sources are Ernst Haas, *Beyond the Nation State*; Richard Rosecrance, *Action and Reaction in World Politics*; Morton Kaplan, *System and Process in International Politics*; Walter Buckley, *Sociology and Modern Systems Theory*; and, of course, the inevitable Karl W. Deutsch, for whose work see the selection entitled "Towards a Cybernetic Model of Man and Society," on view in Walter Buckley, ed., *Modern Systems Research for the Behavioral Scientist* (Chicago: Aldine, 1968). The influence of Figure 3.4 and its ilk is growing, I suspect; the number of papers in political science that now commence with the avowal that society is simply an open feedback system has rocketed toward the higher cardinalities.

what Professor Easton has called "his classical critics."[11] It is a book that has had a wide and only recently waning influence; in an obscure way, only fitfully obvious to outsiders, its appearance marked a change in the character of political science, from what now seems an anachronistic concern for the stable institutions and legal structures of political societies to the relatively modern zeal for analyses that see things in their systematic aspects. Those solid institutions now appear as an iridescent shimmer over a newly found object made up of inputs, outputs, and withinputs, and requiring for *its* explication not the solid, serious, and splendidly dull tools of routine political craftsmanship but the distinctly modern and quasi-magical apparatus of cybernetics, information theory, and communication theory—subjects which when pressed to the purposes of political analysis once seemed technically forbidding and vaguely impalpable. From a point of view fixed *within* political science, these have become *affable* disciplines.

Professor Easton's own introduction to the affable disciplines has been anything but clear: there is much in his style of prose composition that calls cotton wool to mind. In any case, Professor Easton is not an author of passionate scrupulousness: under "system, meaning of," in the index to *A Systems Analysis of Political Life*, there is only one entry. But what we get there is no firm and unwavering explanation of the meaning of systemhood, but a kind of anecdotal account of what Easton has in mind when he argues that it is to the concept of a system that political scientists must repair in order to get a sense of the interconnections beneath the yin and yang of things.[12]

"At the outset," Professor Easton writes, "a system was defined as any set of variables regardless of the degree of interrelationship between them." This brings the systems down simply to the sets, but the remark is saved from vacuousness by a trio of qualifications: systems are situated in a specific environment; systems are "open" in the sense that "by [their] very nature as social system[s] . . . they must be interpreted as lying exposed to influences deriving from the other systems in which empirically they have been embedded"; and finally, systems are *adaptive* in that they have the capacity to respond to disturbances in characteristic ways. Much discussion follows this rather bald avowal, with the conclusion expressed epigrammatically: the *political system*

[11] David Easton, *The Political System* (New York: A. A. Knopf, 1971); also, "Systems Analysis and Its Classical Critics," *The Political Science Reviewer*, Vol. 2 (1974), pp. 269-300.
[12] David Easton, *A Systems Analysis of Political Life* (New York: John Wiley & Sons, 1965), p. 18.

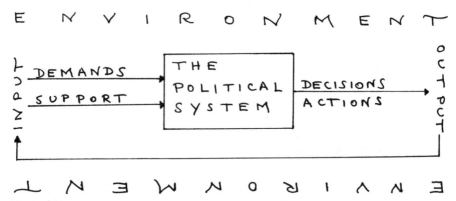

Figure 3.5
The political system under the aspect of Easton. After David Easton, *A Systems Analysis of Political Life* (New York: John Wiley & Sons, 1965), p. 32.

comprises those "*interactions* through which values are authoritatively allocated in society."[13]

An illustration (Figure 3.5) is added to clarify the situation. It recalls Figure 3.4 remarkably: the chief burden of Professor Easton's message is that the political system is in fact a control system, or, at any rate, something very much like a control system.

To the theoretician comes the question: What precisely do you mean? Specificity is the theoretician's curse. Correspondingly, for the shnorfle-suffused critic there is a happy formal fork: either *The Political System* has some quite specific mathematical content or it does not. If not, it has no business using essentially technical terminology; if it does, clobber it on the general grounds that the mathematical models invoked, however indirectly, are qualitatively infelicitous.

This is the Argument from Easton.

Mathematical systems theory beckons the social scientist (*Homo blaatus*). E. Chalmers Ellis sounds a familiar note:

It is one of the not to be underestimated achievements of the modern science of feedback theory that we have come to see, albeit in outline and with many of the not yet revealed details obscure, that market relations are operationally to be defined in terms of such mathematically precise concepts as feedback and feedback loop, where what is being manipulated is not some physical entity, but that mysterious yet all pervasive quantity that mathematicians call information.[14]

[13] Ibid. Notice that by "adaptive" Easton does *not* mean what control theoreticians mean; I discuss this point below.
[14] E. Chalmers Ellis, *The Feedback Structure of the Commodity Market in Frozen Pork Bellies, Hog Futures and Sorghum Grains*, unpublished Ph.D. dissertation, University of Indiana, 1957, p. 31.

But if mathematical specificity is what counts, few if any political systems theorists would survive a serious application of the argument; for most, a commitment to systems theory involves sheer linguistic enthusiasm. In Morton Kaplan's *System and Process in International Politics*, for example, there is talk of feedback, input, and output; it is all done in high spirits, the language is used metaphorically, and there is no suggestion whatsoever that beyond the loose, informal, and disorganized *bavardage* there stands a set of precise mathematical concepts.[15]

In the same vein, Professor Dror writes that he means to "utilize a general systems theory framework of the relations between public policy making and behavioral sciences in order to identify changes required for improved utilization of behavioral sciences for better policy making." Such a theory sees "public policy as an output of the public policy making system, and an input into various target systems."[16] A politician speaks: he whispers, let us say, privileged bits of information to an especially juicy campaign contributor. Looking toward Dror we might say that here we have *output*—the politician's word in MacSwine's ear—and also *input*—MacSwine getting the word.

This is systems theory, then, even general systems theory, but not *mathematical* systems theory, since quite plainly no mathematical concepts are associated with the vocabulary of systems engineering. Against this innocent appropriation I urge no objection save to remark that strictly on aesthetic grounds the traditional vocabulary of political science is preferable to a vocabulary derived from the theory of filters or rheostats.[17]

For the most part, political theorists fall easy victim to the argument from Easton. To turn to Professor Easton is to see the argument self-

---

[15] Morton Kaplan, *System and Process in International Politics* (New York: John Wiley & Sons, 1957).

[16] Y. Dror, "A General Systems Approach to Uses of Behavioral Sciences for Better Policy Making," in E. O. Attinger, ed., *Global Systems Dynamics* (New York: John Wiley & Sons, 1970).

[17] If Kaplan, Deutsch, Easton, and others exhibit the expropriation of the systems engineering vocabulary without its content, Cortes, Przeworski, and Sprague, in their careful volume, *Systems Analysis for Social Scientists* (New York: John Wiley & Sons, 1974), surely demonstrate the orthogonal vice: mathematical systems theory without application. Theirs is an introduction to the methods of the classical tradition; step by unhurried step they take the innocent through the standard topics: inputs and outputs appear on p. 48, the Laplace transform on p. 315; points in between take up the slack. What is missing is anything like a convincing argument that these methods have any more relevance to political and social science than, say, the methods of cohomology theory, algebraic k-theory, or the theory of infinite-dimensional Hilbert spaces.

applied: there is nothing in the entire body of his work—I have measured this carefully—that suggests even a nodding responsiveness to the need, when using a technical language, to fatten it with technical content.

Still, to clinch the argument simply by chasing poor Easton down one fork seems poor sportsmanship. A truly satisfying overall attack would make the double point:

1. The vocabulary is empty at its face.
2. When rigorously, if imaginatively, specified, it turns out to be inappropriate as well.

That political life to *some* extent embodies mechanisms of control is undeniable. The decision to restrict political research to such structures, like any theoretical commitment, must be judged by the fruitfulness of the insights that it yields. I have never been persuaded of the overall importance of *control* itself as a conceptual item in political science, but control theory is what one gets in Easton. I take as a purely polemical assumption the view that Figure 3.5 is really a stand-in for Figure 3.4: behind them both is the classical tradition. Mushing on with the argument from Easton will require first that the mathematical commitments of the classical tradition be captured in a space of some reasonable dimension, and second that the points making up that space be compared with some specification of the mathematical properties that are likely to characterize social and political life.

I have in mind a comparative exercise of the sort carried out by Professor Grodins when he remarks, in the context of a discussion of biological control theory, that "the classical theory . . . is applicable only to simple linear and time invariant systems. However, it is becoming increasingly clear that biological systems are complex, non-linear and time variant."[18] Professor Anne has followed this up with a table that makes the point-by-point case (Table 3.1). From it he concludes, quite rightly, that applying classical control theory to biological problems is an exercise in misapplied force, like hammering a screw into wood.[19]

Anne's chart lists nine points by which a comparison between biological and classical control theory might be made. I am not sure what he means in the last three, so I have devised my own somewhat more

---

[18] F. S. Grodins, "Similarities and Differences Between Control Systems in Engineering and Biology," in Attinger, *Global Systems Dynamics*, pp. 2-6.
[19] For another view, see J. H. Milsum, *Biological Control Systems Analysis* (New York: McGraw-Hill, 1966), a volume from which one can learn much control theory, little biology.

Table 3.1
Characteristics of control systems in engineering and biology

	Physical	Biological
1. System control	Lumped	Distributive
2. Organization	Simple	Complex
3. Dimensionality	Low	High
4. Linearity	Linear	Nonlinear
5. Time dependence	Invariant	Variant
6. Methodology	Synthesis/Analysis	Analysis/Synthesis
7. Nature	Goal-seeking by deterministic and stochastic processes	Goal-seeking by evolution and adaptation
8. Interactions of subsystems	Fixed	Varying
9. Redundancy in design	None	Much

Source: A. Anne, "Comments on a Paper by F. S. Grodins," in E. O. Attinger, ed., *Global Systems Dynamics* (New York: John Wiley & Sons, 1970), p. 7.

extensive grid (Table 3.2), and it is on this basis that I forge on against Easton. Below, I discuss the three specific points of greatest importance.

Dimensions and differential equations[20]

Classical systems analysis is a subject inseparable from the theory of ordinary differential equations. There are, to be sure, finite-difference equations that can go proxy for ordinary differential equations, but the experienced mathematician passes from one to another easily. So the question remains whether ordinary differential equations are suitable instruments for modeling social and political processes. If not, this amounts to an a priori rejection of Figure 3.4, whatever the merits of control theory as a mirror of the political, since Figure 3.4 presupposes the ready availability of dynamical models phrased as differential equations.

Given the extraordinarily broad character of the question, no decisive answers are possible. But there are infelicities in the application of ordinary differential equations that should give the theorist pause. I mention two: the first deals broadly with the continuity assumptions that are a fixed and ineradicable part of the apparatus of ordinary differen-

---

[20] The question of the degree to which ordinary differential equations are suitable to the social sciences was discussed in a slightly different context toward the end of Chapter 2. I pursue this same theme in a paper, "Mathematical Models of the World," *Synthese*, Vol. 31, No. 2 (1975).

Table 3.2
Comparative characteristics of the classical tradition of mathematical systems theory and social systems (the elevenfold way)

	The Classical Tradition	Social Systems
1. Dynamics	Ordinary differential equations	Often unknown
2. Dimensionality	Low (input–output)	Very, very high
3. Linearity	Linear	Nonlinear
4. Coefficients	Constant	Variable
5. Time dependence	Autonomous	Nonautonomous
6. Domain	Frequency (complex)	Time (real)
7. Mode of analysis	Laplace transform and transfer function	Dependent on the particular dynamical structure
8. Parameters	Lumped	Distributed
9. Regulation/Control	Output feedback/Fixed plant	Adaptive control
10. Typical problems	Stability; error; compensation design	Nature of the dynamical structure
11. Nature of control	Regulator plus input–output system	Feedback by system of simultaneous equations

tial equations themselves; the second, with the gluey and mathematically intractable complexities that arise when differential equations swell dimensionally.

Differential equations are tied to continuous and differentiable functions.[21] Counting this an objection to differential models means making the case that social and political phenomena are discontinuous, but even then the machinery of ordinary differential equations, suitably handled, can reflect discontinuous phenomena from within. The study of singularities, for example, which figures so prominently in general

---

[21] Quite different objections to the use of ordinary differential equations in the context of classical control theory and the theory of linear systems arise when one remembers that parameters specifying a system are generally presumed to remain constant over space. For simple systems, this is reasonable enough: purists will note, however, that the acceleration of an object *need* not be uniformly distributed to its parts. To specify such systems fully, relevant parameters would have to be defined in a way that specifies variations in force as functions of variations in space. Here *partial* differential equations are needed. Systems in which it is safe to stick with ordinary differential equations are *lumped*; all others, *distributed*. Distributed parameters become important under conditions in which inputs and outputs are specified over vast areas—just what one would expect in the social and political sciences.

relativity, involves points in the solution set to a system of differential equations where the solution itself is undefined.

Bifurcations occur also in the qualitative theory: points arise that separate zones of radically discontinuous sets of solutions. Given these resources, differential models may very well reflect discontinuous processes. But their use in such contexts is mathematically unnatural, and purists will suspect that ordinary differential equations are *generically* unsuited as recording instruments for unstable processes. The method, Professor Sussman remarks,

does not lend itself very easily to the study of situations which essentially involve objects which remain unchanged for long intervals of time and then experience abrupt changes. The state of a system described by ordinary differential equations is either always changing or, if it is constant for a very short time, it remains constant forever, unless it is disturbed by an agent extraneous to the system.[22]

Even without a detailed argument, it is clear that many social and political situations undergo processes of just this sort, as when a relatively minor incident triggers a major political conflagration or a long period of equilibrium comes to a sudden halt.

The instability of social life, then, suggests certain reservations with respect to ordinary differential equations. Its sheer intrinsic complexity suggests others still. By "complexity" I mean specifically a measure of the number of variables that interact crucially with a variable of interest. Of two systems

$$\frac{dx_i}{dt} = f_i(x_1, \dot{x}_2, \ldots, x_n), \qquad i = 1, 2, \ldots, n, \tag{3.20a}$$

$$\frac{dz_j}{dt} = h_j(z_1, z_2, \ldots, z_p), \qquad j = 1, 2, \ldots, p, \tag{3.20b}$$

the first is more complex than the second if $n > p$.

It was Richard Bellman, I believe, who introduced the happy phrase, "the curse of dimensionality," as a description of the computational intractability of large systems. It is a kind of complexity that the clas-

---

[22] H. J. Sussman, "Catastrophe Theory," *Synthese*, Vol. 31, No. 2 (1975), p. 4. Short of Thom's book itself, this is the best introduction to catastrophe theory. For a study of singularities, see S. W. Hawkins and C. F. R. Ellis, *The Large Scale Structure of Space-Time* (London: Cambridge University Press, 1973), pp. 256–299.

sical tradition steadfastly eschews: throughout there is an obdurate preference for the simple scheme that takes a single input to a single output. This alone is reason enough to reject Figure 3.4 as a workable model for political processes; at the very least one needs a transmutation of $x(t)$ to $\{x_1(t), x_2(t), \ldots, x_n(t)\}$, where the $x_i(t)$ are all components of a *vector*. This one in fact can achieve in the state-space approach to control theory (discussed below), but the need for complex relationships nevertheless leads to characteristic inadequacies "seen whenever one has to deal with a large number of variables which cannot easily be aggregated into a small number of macroscopic ones."[23] Failures in aggregation lead to differential difficulties largely because of the behavior of ordinary differential equations under what Professor Sussman calls "pushing forward":

Indeed, suppose we have a system [such as (3.20a)] with n very large and suppose that there is a small number m of important variables y_1, $y_2, \ldots, y_m$ which are functions of the x_i, namely,

$$y_k = g_k(x_1, x_2, \ldots, x_n), \qquad k = 1, 2, \ldots, m \tag{3.21}$$

Suppose that the y_k are the only variables that really matter to us. It would then be desirable to use (3.20a) to get a similar system of ordinary differential equations for the y_k. Now, it is easy to see that, *in general, such a system will not exist.* For instance, (a) the future behavior of the y_k will not necessarily be determined by their values y_k^0 at a particular time t_0, as would be the case if the y_k evolved according to ordinary differential equations.[24]

The conditions that Sussman indicts are just the ones that might obtain if a refitted version of Figure 3.4 were to hold for social systems. Realistic inputs to plants governing social systems must be multidimensional; variables over which the functions range are seldom likely to be perfectly aggregable. The very format of the input-output mode demands that what is of chief interest is the relationship of the plant to those output functions y_k that are connected to the differential equations themselves. In these circumstances one would not expect to see output functions *regularly* related to the plant's dynamical structure; worse, the traditional manner of dealing with these difficulties is not feasible when one passes to very large systems:

---

[23] Sussman, "Catastrophe Theory," p. 4. Sussman's own example has to do with problems, by now notorious, of aggregating utility functions into a social-welfare function.
[24] Ibid.

Now, systems exhibiting [this] feature are far from a mathematical novelty. For instance, if the y_k are the coordinates of the position of a moving particle, then (a) holds. The traditional way out of this difficulty is approximately three centuries old, and it consists of adding new state variables, such as the components of the moving particle's velocity vector. However, in recent years we have witnessed developments for which this traditional way does not work. In the social sciences, in biology, in geology, the number of variables becomes too large and the models become impossible to handle.[25]

The theory of ordinary differential equations is one of the glories of the mathematical method; together with the theory of partial differential equations, it comprises the chief mathematical instrument in theoretical physics. But mathematicians working in biology, in the social sciences, and in meteorology have come to see that the simple, straightforward, and standard methods of mathematical physics do not work with the same perfection in the analysis of, say, the passage of fiscal legislation through the lower chamber of the house in Swoboda County, Illinois, as they do in the analysis of the movement of point masses in a field of force. Writers who argue the deficiencies of these methods do not generally do so with only a blank to commend in their place, and there are, in fact, differential theories without differential equations. Smale and Thom, for example, have written about such methods.[26]

Linearity

Complexity begets nonlinearity. But linear theory is where the theo-

---

[25] Ibid. For a description of the use of generalized coordinates in classical dynamics, see Herbert Goldstein, *Classical Mechanics* (Reading, Mass.: Addison-Wesley, 1950), pp. 10-14.

[26] See, for example, Stephen Smale's "Global Analysis and Economics," reprinted in *Synthese*, Vol. 31, No. 2 (1975). It is catastrophe theory, of course, that has generated excitement recently. Thom's book has been translated into English by W. D. Fowler under the title *Structural Stability and Morphology* (New York: W. A. Benjamin, 1975). J. Guckenheimer has written about catastrophe theory in an article entitled "Catastrophes and Partial Differential Equations," in the *American Institute Fourier Grenoble*, Vol. 23, No. 2 (1973), pp. 31-59, and E. C. Zeeman has provided a first important application of Thom's theory in a paper entitled "Differential Equations for the Heartbeat and Nerve Impulse," in M. M. Peixoto, ed., *Dynamical Systems* (New York: Academic Press, 1973). Thom's theory is devoted to phenomena of evolutionary instability: slow passage through a space of controls is associated with sudden, indeed catastrophic, instabilities in an associated parameters space. The chief object of the theory is the classification of the singularities that arise on the catastrophe boundary. In what must be reckoned an extraordinary convergence of interests, Thompson and Hunt have developed a mathe-

rems are. Buridan's ass perished between two such choices.

In the face of nonlinearities, the prevailing tactic among economists, at least, has been to linearize wherever possible, and where not, to linearize anyway. This gives the resulting theories power, though often at the expense of qualitative realism. The social sciences as a whole still share with nineteenth-century physics the view that the universe is inherently linear; deviations are discounted much as Kepler discounted the ellipse. The physical sciences, however, have long since accepted nonlinear phenomena as ineradicable. Ironically enough, this is especially true in the systems sciences: only nonlinear mathematics describes oscillatory phenomena of the sort studied, say, by van der Pol in the analysis of electronic amplifiers. This work, merged with the great work of Poincaré, Liapunov, and Hill, stands sniffishly apart from the vast work in strictly linear theory.[27]

Linearity as an idea is best appreciated in the context of abstract vector spaces. Such spaces are also called *linear spaces*, and the terminology reflects the degree to which ordinary and intuitive concepts of linearity have been impressed into an imposing mathematical creation. Briefly, an abstract vector space consists of a set of vectors $V = \{x, y, z, \ldots\}$ taken together with a set of scalars $F = [a, b, c, \ldots]$ such that vector addition and scalar multiplication are both well-defined. A first set of axioms guarantees that vector addition is commutative and additive and that, moreover, an additive inverse and an identity element exist:

1. $x + y = y + x$
2. $x + (y + z) = (x + y) + z$
3. $x + 0 = x$, for all x in V for a unique zero vector 0
4. for every x in V there exists an inverse, $-x$, such that $x + (-x) = 0$; the second that scalar multiplication is well-defined:
5. $1 \cdot x = x$
6. $(ab)x = a(bx)$

---

matical scheme from the perspective of optimization engineering techniques that shares something of the structure of catastrophe theory itself and in any case is devoted to the same class of problems; see J. M. T. Thompson and G. W. Hunt, *A General Theory of Elastic Stability* (New York: John Wiley & Sons, 1973).

[27] See, for example, S. Lefschetz, *Stability of Non-Linear Control Systems* (New York: Academic Press, 1965), or J. K. Aggarwal, *Notes on Nonlinear Systems* (New York: Van Nostrand Reinhold, 1972), or Austin Blaquière, *Non-Linear System Analysis* (New York: Academic Press, 1966). A text that has recently caused some excitement is S. M. Goldfeld and R. E. Quandt, *Nonlinear Methods in Econometrics* (Atlantic Highlands, N.J.: Humanities Press, 1972).

7. $a(\mathbf{x} + \mathbf{y}) = a\mathbf{x} + a\mathbf{y}$
8. $(a + b)\mathbf{x} = a\mathbf{x} + b\mathbf{x}$.

These axioms define a vector space $\mathbf{V}$ over the field F: such spaces can also be described in group-theoretic terms as Abelian groups with operators. Abstract vector spaces lie athwart any number of rich algebraic concepts, such as modules, rings, and algebras; Euclidean space itself is derived from structures simple as a vector space impressed with norm and inner product.

Basic objects in linear theory are the *transformations* or *homomorphisms H* from vector spaces to vector spaces. It is their algebraic structure that becomes the chief subject of linear algebra. Linear transformations satisfy axioms of both superposition and homogeneity: i.e., if $\mathbf{V}$ and $\mathbf{W}$ are vector spaces over F, a homomorphism or linear transformation from $\mathbf{V}$ into $\mathbf{W}$ is defined as a function H such that
1. $H(\mathbf{x} + \mathbf{y}) = H(\mathbf{x}) + H(\mathbf{y})$
2. $H(a\mathbf{x}) = aH(\mathbf{x})$
for all $\mathbf{x}$, $\mathbf{y}$ in $\mathbf{V}$ and a in F. It is a trivial consequence of the definition that linear transformations preserve linear combinations: if $x_1, x_2, \ldots, x_n$ are vectors in $\mathbf{V}$, and $c_1, c_2, \ldots, c_n$ are scalars, then

$$H(c_1 x_1 + c_2 x_2 + \ldots + c_n x_n) = c_1 H(x_1) + c_2 H(x_2) + \ldots + c_n H(x_n).$$

It is another consequence of the definitions that the set H^* of all linear transformations or homomorphisms between two vector spaces $\mathbf{V}$ and $\mathbf{W}$, when taken together with operations of functional addition and functional scalar multiplication, itself inherits the structure of an abstract vector space.

In systems theory, the transformations L taking inputs $x(t)$ to outputs $y(t)$ are linear if

$$L[x(t)] = y(t)$$

is a linear mapping in the straightforward sense described above. It is not my purpose here to develop the appropriate notions of systems linearity in their full generality: in keeping with my stress on the dynamical character of the classical tradition, I accept a system as linear if the differential equations animating the plant may be written in the form

$$\frac{dx_i}{dt} = \sum_{j=1}^{n} a_{ij}(t)x_j + b_i(t), \qquad i = 1, 2, \ldots, n. \tag{3.22}$$

This is not entirely satisfactory since some linear systems are not specifically governed by equations of this form, but it will do for the present discussion.

Linearity is often obscured in discussions of classical control theory because transform and transfer methods obliterate the structural characteristics of the dynamical equations. So there is something revealing about treating the differential equations from a strictly differential point of view, with no thought whatsoever for systemwide properties or the relationship of inputs to outputs.

The simplest interesting case is the homogeneous single-variable equation

$$\frac{dx}{dt} = ax, \qquad\qquad (3.23)$$

whose solutions are

$$x = ke^{at}, \quad k \text{ an arbitrary constant.} \qquad\qquad (3.24)$$

Linear algebra comes into play on this elementary level with a vengeance: the solutions to (3.23) form a one-dimensional linear space: any linear combination of solutions is itself a solution.

The inhomogeneous equation

$$\frac{dx}{dt} = a(t)x + b(t) \qquad\qquad (3.25)$$

represents one step up in complexity: complete solutions are given by Leibnitz's formula,

$$x = e^{\int a(t)dt} \left(\int e^{-\int a(t)dt} b(t)\, dt + c \right). \qquad\qquad (3.26)$$

The interpenetration of linear algebra and differential theory reappears in linear *systems* of first-order ordinary differential equations. For the homogeneous case

$$\frac{dx_i}{dt} = \sum_{j=1}^{n} a_{ij}(t)\, x_j, \quad i = 1, 2, \ldots, n, \qquad\qquad (3.27)$$

any linear combination of solutions is again a solution, and the set of solutions consequently takes on the structure of an n-dimensional linear space over the complex field.

The linear structure of the homogeneous case makes possible a solution to the inhomogeneous system (3.22). Indeed, the inhomogeneous case gets solved, basically, by means of superimposed solutions derived from the homogeneous system. For systems with variable parameters, no easy solutions inevitably suggest themselves; only systems with constant coefficients are completely manageable. Still, if enough solutions to (3.27) are available, the general case comes as an oblation from linear algebra.

Nonlinear equations have none of this structural luminosity: solutions are determined by various overlapping methods. Famous equations appear, one after another, like ships in the night: Lienard's equation, van der Pol's equations, Abel's equations, Duffing's equation, the Volterra-Lotka equations. Esoteric techniques materialize: perturbation techniques; techniques involving small parameters; the method of Krylov-Bogliubov; the method of Lighthill and Temple; Goodman and Sargent's technique; the method of collocation; the Ritz-Galerkin method; the method of Lie series; the method of complex convolution; the method involving the celebrated Taylor-Cauchy transform. Nothing suggests the sheer organizational genius of the linear theory.

Why, then, hesitate at the use of linear methods in the social sciences when from a mathematical point of view they behave with such superb and unfaltering regularity?

One line to which I give a respectful nod comes to the bald avowal that many interesting political or social processes are obviously not linear. The easiest way to generate nonlinearities is to imagine a state variable itself invoked as an operator—as when the computation of a derivative involves the multiplication of a pair of state variables. This is the mechanism behind the mathematics of diffusions, infusions, the peddling of influence, the spread of bacteria across the surface of a plate, contagions, rumors, epidemics—in short, just those processes one would instinctively suppose sustain the social and political world.

This is really an argument by inappropriateness: given the mathematics that one suspects informs the Social, strictly linear methods will not do.

But there is also the evidence of research in the physical sciences and in engineering itself. The view that nonlinearities are disguised linearities has not been current in the physical sciences for half a century or so. The reason plainly is the discovery and analysis of physical phenomena that are *intrinsically* nonlinear. Areas of severe nonlinearities include theories of subharmonic resonance, large-deflection theory, the theory of elastic finite deformations, the theory of incompressible fluids (indeed, hydrodynamics generally), and theories of turbulence and wave flow.[28] Here nonlinear systems generate qualitative properties that *cannot* in the nature of things arise in correspondingly linear systems.

Limit cycles provide a case in point. Consider the system of nonlinear equations,

$$\frac{dx}{dt} = y + x(1 - x^2 - y^2), \qquad \frac{dy}{dt} = -x + (1 - x^2 - y^2), \qquad (3.28a)$$

as compared to the linear harmonic oscillator equations

$$\frac{dx}{dt} = y, \qquad \frac{dy}{dt} = -x. \qquad (3.28b)$$

Introduce damping forces in (3.28b) and the form of the solutions changes drastically: (3.28b) undamped exhibits solutions in periodic form; damped, solutions approach zero or infinity as the damping coefficient is either greater or less than zero. Oscillating solutions to (3.28a) have quite a different character: in fact, the amplitude of oscillation of a negatively damped, stable nonlinear oscillator may well tend to a finite limit. Such are the limit cycles (Figure 3.6). They represent, in effect, isolated orbits in the phase plane. Under plausible conditions, it is possible to prove for equations of the Lienard form, of which (3.28a) is an example, that any nontrivial solution is either a limit cycle or a spiral that tends to a limit cycle as $t \to \infty$. Moreover, limit cycles *cannot* arise in the context of conservative linear systems.

The qualitative difference between the systems is striking: plainly, certain physical phenomena cannot be faithfully modeled by linear mathematical methods.[29]

---

[28] See, for example, I. Stakgold, D. Joseph, and D. Sattinger, eds., *Nonlinear Problems in the Physical Sciences and Biology* (New York: Springer-Verlag, 1973).

[29] Periodic processes, from simple clocks to biological regulators, seem appropriately modeled differentially by nonlinear equations with stable limit cycles. And this fact has not been lost on biologists:

The existence of multiple limit cycles, like the existence of multiple stationary states, suggests a basis for explaining the myriad of quasi-discontinuous events observed in biological systems. . . . The insensitivity that biological limit cycles would have to a wide range of initial conditions and perturbations is particularly appealing in the context of biological clocks, evolutionary biology, macromolecular

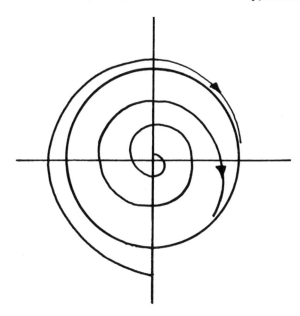

Figure 3.6
A sketch of two solutions to (3.27), showing the limit cycle. (The equations can be solved fairly easily if they are first transformed to polar coordinates.)

A concatenation of circumstances, then, makes linearity a point of exquisite sensitivity in the appraisal of the usefulness, for the political or social sciences, of such models as elementary systems theory affords. In the first place, it is plain that the classical objects studied in systems theory are plants made pliant by the most obvious of linear assumptions. But it is plain too that realistic accounts of social and political processes are apt to be nonlinear. And in the third and, to avoid too depressing a list, final place, it is mathematically and physically clear that some processes and phenomena are nonlinear in a way that is inescapably resistant to easy linearization methods. This means, of course, that the architectural creations of the linear differential theories cannot be brought to bear. Of course, a considerable body of work has been

synthesis, and the like. Mechanisms for which multiple limit cycles and parametric variation are possible should remain high on the list of candidates for the explanation of quasi-discontinuous events in biological systems. [Charles F. Walter, "Kinetic and Thermodynamic Aspects of Biological and Biochemical Control Mechanisms," in E. Kun and S. Grisolia, eds., *Biochemical Regulatory Mechanisms in Eukaryotic Cells* (New York: Wiley-Interscience, 1972), p. 373]

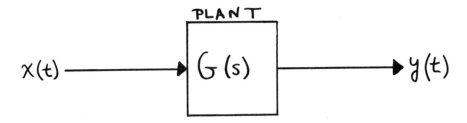

Figure 3.7
A simple system.

devoted to nonlinear systems theory, and to such work the social and
political scientists might turn when confronted with the charge that
social and political life exhibit alarming nonlinearities. I do not mean
for a moment to suggest that nonlinearity moves a subject beyond
mathematical analysis. But the role of linearity, properly appreciated,
makes it quite difficult to accept as central for the social and political
sciences engineering models that are *basic* in the analysis of linear systems; in rejecting these models, social and political scientists are rejecting the most intuitively attractive and most thoroughly understood
of engineering objects. And what they get in exchange is a set of structures far more poorly understood. It is possible for a political scientist—
MacEaston, say—to accept as a schema the system set out in Figure 3.5
while rejecting the standard linear account that describes it. But then
he is obliged to supply the missing mathematics.

Adaptive control[30]

Associated with any plant is a set of parameters: for a plant governed
by ordinary differential equations, these will just be the coefficients
of the equations themselves. Figure 3.1 shows the basic process: inputs
passing to outputs by means of some specified operations in the plant
itself. In Figure 3.7 the dynamics are fully specified and the system is
described by means of block diagrams. This becomes a control system

---

[30] For a somewhat out of date but valuable survey of adaptive control, see E. Mish-
kin and L. Braun, eds., *Adaptive Control Systems* (New York: McGraw-Hill, 1961);
a more recent collection is Rufus Oldenburger's *Optimal and Self-Optimizing Con-
trol* (Cambridge, Mass.: The MIT Press, 1966). See also the survey paper by P. H.
Hammond and N. W. Rees, "Self-Optimizing and Adaptive Control Systems," in
R. H. Macmillan, ed., *Progress in Control Engineering*, Vol. 3 (London: Heywood
Books, 1966). For the modern algebraic approach, see Part 4 of Kalman, Falb, and
Arbib, *Topics in Mathematical Systems Theory*.

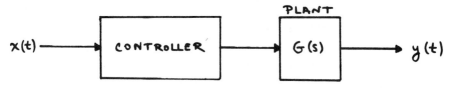

Figure 3.8
A simple system with a controller added.

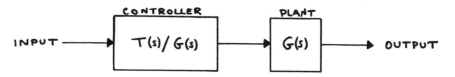

Figure 3.9
Example of the use of a controller to modify a system.

on the introduction of a controller, as in Figure 3.8. The basic problem is to construct a controller that will direct outputs onto preassigned paths. If the dynamics of the process are perfectly known, specifying the controller is trivial: invert the original dynamics and replace them with the preferred dynamics: if $T(s)$ is required, use $T(s)/G(s)$ as the controller, as in Figure 3.9. There is no need for feedback in the face of perfect information.

There are two circumstances in which this simple solution is not possible. The first occurs when the dynamics of the plant are known but parameters may vary, so that perfect control is possible only indirectly. This is pretty much the case envisioned by the classical tradition. Or again, the plant itself may be subject to disturbances, or the input may vary in ways that cannot be foreseen. In these circumstances, control can be achieved only by feedback.

As against these benign disturbances is the possibility that plant parameters may vary widely or fluctuate subject to no formal predictions. The classical tradition is useless in such cases, and one needs instead the techniques of *adaptive control theory*. The simple fact that in adaptive control theory the system itself turns slippery makes for profound mathematical difficulties. The most important arise when attempts are made to discover just how the plant's parameters are changing. This is the *identification problem*: at its most extreme, nothing is known of the dynamics of the plant, and the task of regulation begins with an effort to assemble a transfer function strictly on the basis of input and output information. Clearly, a *complete* mathematical solution to the identification problem would be remarkable: given the rules for system identification, should they exist, it remains only to apply them to particular cases in order to determine the functional relationships between any pair of causes and their effects.

The problem of adaptive control, then, is two-staged: identification of process dynamics first, followed by the virtually continuous reconstruction of the plant parameters themselves. This is *feedback by parameters*.

For the case of linear, constant, time-invariant systems, much of the problem of adaptive control is understood. R. E. Kalman synthesized some partial results in 1958, and although his paper is mathematically ungeneral, it gives a good idea of the nature of the problem in the classical setting.[31]

In Kalman's view, adaptive control is a problem of three parts: (1) to determine the dynamical characteristics of a plant; (2) to determine the characteristics required of the controller in order to achieve design specifications; and (3) to construct a controller with the appropriate characteristics. I scout only his solutions to (1), treating (2) and (3) as peripheral matters of design.

Begin with the block diagram of Figure 3.7; $x(t)$ and $y(t)$ are related by a convolution integral:

$$y(t) = \int_{-\infty}^{t} g(t - \tau)\, x(\tau)\, d\tau, \qquad (3.29)$$

where $g(t)$ is the unit-impulse response of the process. The identification problem is soluble if (3.29) is; failing that—approximate. If both input and output are sampled discretely, (3.29) becomes

$$y_k = \sum_{i=-\infty}^{k} g_{k-l}\, x_l, \qquad (3.30)$$

where y_k and g_k are values of $y(t)$ and $g(t)$ at the kth sampling moment. Process dynamics are now given explicitly by the infinite set of numbers g_k. This involves an infinite number of corresponding algebraic equations, but under stable conditions the sampled points converge.

A more efficient system of solution is obtained by defining a z transform or pulse-transfer function

---

[31] R. E. Kalman, "Design of a Self-Optimizing Control System," *American Society of Engineers Transactions*, Vol. 80 (February 1958), pp. 468–478.

$$G(z) = \frac{y(z)}{x(z)} = \frac{a_0 + a_1 z^{-1} + \ldots + a_n z^{-n}}{1 + b_1 z^{-1} + \ldots + b_n z^{-n}}, \tag{3.31}$$

which plays the same role in the analysis of linear sampled-data systems as the pure transfer function plays in the analysis of continuous systems.

Inputs and outputs are related by a difference equation:

$$y_k + b_1 y_{k-1} + \ldots + b_n y_{k-n} = a_0 x_k + a_1 x_{k-1} + a_2 x_{k-2}$$
$$+ \ldots + a_n x_{k-n} \tag{3.32a}$$

or, solving for y_k,

$$y_k = a_1 x_{k-1} + a_2 x_{k-2} + \ldots + a_n x_{k-n} - b_1 y_{k-1} - \ldots - b_n y_{k-n}. \tag{3.32b}$$

Substituting into (3.32b) the values $a_i(N)$ and $b_i(N)$ of the coefficients at the nth sampling moment gives the past values of y_k, which we dub $y_k(N)$:

$$y_k(N) = a_1(N)x_{k-1} + a_2(N)x_{k-2} + \ldots + a_n(N)x_{k-n} - b_1(N)y_{k-1}$$
$$- \ldots - b_n(N)y_{k-n}, \qquad k = 0, 1, 2, \ldots, N, \tag{3.33}$$

subject to the condition that the values of the $a_i(N)$ and $b_i(N)$ are chosen so as to minimize

$$\frac{1}{N} \sum_{k=0}^{N} [y_k - y_k(N)]^2 .$$

Thus the coefficients $a_i(N)$ and $b_i(N)$ identify the plant at the nth sampling moment; the identification problem is solved when the procedure is repeated for each n.[32]

With the solution of the identification problem in the form of explicit equations specifying the values of the coefficients, only the construction of the controller remains. This will, in general, depend on the

---

[32] See John E. Gibson, *Nonlinear Automatic Control* (New York: McGraw-Hill, 1963), pp. 516–517, from which the above account of the Kalman pulse-transfer technique was taken.

performance criteria involved: given a specific constraint, the controller will be geared to the construction of an appropriate forcing function.

Kalman's solution to the identification problem is by now a mathematical staple: abstract realization theory has been overcome by algebraists. For an up-to-date appreciation, one must consider Ho's algorithm and other state-space techniques.

I have gone through this exercise for pedagogical purposes. In biology and in the social sciences, one thinks of adaptive control as natural: the mechanisms by which social life is regulated, once one excludes certain artificial cases, inevitably involve adaptive processes, with the machinery of regulation employed in a continuous redefinition of the regulated process itself.

This by itself is a reason to count the lack of adaptive control as an episode in the argument from Easton. Figure 3.4 is a control-theoretic object: it exhibits feedback but not adaptive control since it sports no mechanisms for the identification of dynamical processes.

This leaves open the response, of course, that if adaptive control is needed for the successful modeling of political processes, this involves only an attainable enrichment at Figure 3.4. But this is to dismiss the conditions that make for an easy solution to the identification problem—namely, that the underlying systems be presumed linear, invariant, and stable. Linearity is hardly a suitable constraint; invariance betokens a set of structural relations that do not change with time; and stability suggests anything but the processes that work their way through social and political life.

Reflections on the Theory of Linear Systems: State Variables and the Generalized Argument from Easton

An automobile which runs out of fuel will not function.

Oran R. Young[33]

The most notable limitation of the classical approach is simply its adhesion to (3.6) as the dynamical core to systemhood. This refinement of scope makes possible the relief afforded by the Laplace transform and the resulting transfer function; still, there is something conceptually indecent about the algebraic elimination of the system's true differential structure. Hence the aesthetic motivation for the modern *state-variable* approach to linear systems.

[33] Oran R. Young, *Systems of Political Science* (Englewood Cliffs, N.J.: Prentice-Hall, 1968), p. 22.

Professors Kalman, Falb, and Arbib write that a system is "a structure into which something (matter, energy or information) may be put at certain times and which itself puts out something at certain times."[34] This is a definition much in the spirit of Figure 3.1. What separates the modern from the classical point of view is the analytically satisfying emphasis on the notion of state. Taking a given system Σ to be a three-pronged affair with $u(t)$ representing inputs and $y(t)$ outputs, Kalman, Falb, and Arbib write that

we may not be able to say what $y(t)$ is without knowing more than just what the present input $u(t)$ is. The past history of inputs to the system may have altered Σ. . . . In other words, the output of Σ in general depends on both the present input of Σ and the past history of Σ. . . . We say, therefore, that the present output depends on the state of Σ, and we define the present state of Σ intuitively as that part of the present and past history of Σ which is relevant to the determination of present and future outputs. In other words, we think of the state of Σ as being some (internal) attribute of Σ at the present moment which determines the present output and affects the future outputs. Intuitively, the state may be regarded as a kind of information storage or memory or an accumulation of past causes.

This is an account made deliberately complex in order to support a stock of later theorems. A more intuitive picture of the state variable comes from sloughing off some of the definition's generality.

Fundamentally, there are variables x_i representing the system's states, u_j representing its inputs or controls, and y_k representing its outputs. The states themselves need not be *physically* definable.

The classical theory pegs systemhood to devices that take a single input to a single output—filters, for example, or transducers. The modern theory is a chapter in the history of linear algebra. Kalman, for example, has recently urged a *module-theoretic* construction of the *entire* theory of linear systems. Systems are usually specialized to *abstract vector space*, so that the basic objects over which the theory ranges are vectors and not, as in the classical theory, real numbers. It is just this expansion toward vectors that makes possible the elimination, in the modern theory, of the tacit if pervasive classical requirement that inputs and outputs be singled-valued; multiple-valued inputs can be represented as components of a vector.

The state, in turn, appears as a collection of points defined in the

---

[34] Kalman, Falb, and Arbib, *Topics in Mathematical System Theory*, p. 4. For an introduction to the state-space approach, see Katsuhiko Ogata, *State Space Analysis of Control Systems* (Englewood Cliffs, N.J.: Prentice-Hall, 1965).

n-dimensional vector space. The state-transition function holds sway now as the *solution* to a vector-matrix differential equation:

$$\frac{d\mathbf{x}}{dt} = \mathbf{A}(t)\,\mathbf{x}(t). \tag{3.34}$$

Given boundary conditions at some t_0, we denote the values of the solution vectors at time t by $\Phi(t, t_0)$; thus

$$\frac{d\Phi(t, t_0)}{dt} = \mathbf{A}(t)\,\Phi(t, t_0). \tag{3.35}$$

The transition function, of course, and the transfer function play something of the same role in the modern and the classical theories. Both give expression to the system's overall connection of input to output. But the classical theory is heavily involved in *obliteration* of state-space considerations: the Laplace transform and the mechanism of the transfer function serve to conceal the dynamical system's differential core. Thus the classical theory is really bimodal—in and out. The modern theory, by way of crude contrast, is trimodal—in and out by means of a set of states. (The key points between theories are set out in Table 3.3.)

Differences in power, scope, and generality notwithstanding, in their

Table 3.3
Comparison of the classical and state-variable theories

	Classical	State-Variable
Focus	Input–Output	States
Representation	Differential equations Laplace transform Transfer functions	Differential equations Transition function
Feedback	By means of output	By means of state
Dimensions	Low	High
Linearity	Linear	Linear
Domain	Complex (frequency)	Real
Time-dependence	Autonomous	Autonomous
Coefficients	Constant	Constant
Control-theory design	Scattered results	Unified theory
Control objects	Closed-loop feedback system Control law minimizing error	Closed-loop feedback system State-estimation regulator

fundamentals both classical and modern theories describe systems by means of linear differential equations that are at once autonomous, homogeneous, and constant. The state-variable approach to the analysis of linear systems is the classical tradition revamped with a gain in power and a certain increase in flexibility. But the mathematical limitations that made classical methods a poor choice for the analysis of social and political systems are a part of the modern tradition as well.

A natural question to ask, then, considering the whole of linear systems theory, is whether there is anything especially compelling to the assumption, which runs like a flashing current through so much sociology, political science, and psychology, that systems taken as suitable by the theory of linear systems are suitable also as models, however idealized, for the social and political sciences. I pursue this question under three headings: Feedback, Input-Output Descriptions, and Control Theory.

Feedback

That society is a system often involves *only* the assumption that social systems instantiate mechanisms of feedback. In Keynesian theory, for example, national income Y is usually thought of as the sum of two expenditures: investments A and consumer goods C. C in turn is a linear function of Y; thus, overall,

$$Y = A + cY \quad \text{or} \quad Y = \frac{A}{1-c}. \tag{3.36}$$

The similarity of this expression to "the basic formula of the theory of regulation" suggested to Oskar Lange "an interesting interpretation of the Keynesian multiplier." The interpretation that he has in mind, however, comes to nothing more than an arbitrary *reinterpretation* of economic terminology:

Let us assume that we have a certain system with the "input" . . . $x = A$. The transmittance of the system $S = 1$, which means that investments result in expenditures equal to the level of investment outlays A. There is a feedback relation between the system and the governor with the transmittance $R = c$. After the correction introduced by the governor the overall input of the first system is the activity of the size $Y = A + cY$.

This example, Professor Lange concludes improbably, "confirms what we said in the beginning, namely that the economists deal with problems of regulation and use formulae from the theory of regulation without knowing it."[35]

This is harmless enough, but the concept of feedback *has* had a dangerous influence among social scientists: there is a widely held view, for example, that in feedback the psychologist or philosopher has the means by which an analysis of *goal direction* might be achieved.[36]

The idea that there is something especially significant to the notion of feedback is Norbert Wiener's: his was the hope that some set of concepts might suitably describe both human behavior and the performance of machines in a fashion calculated to explain the first by describing the second.[37] A large-minded project this, ostensibly carried out in the service of a dispassionate materialism; much effort has been spent in its name simply to show that beneath the familiar edges and surfaces of the human body there remains no irreducible minimum that cannot be explained by the way all the rest is organized and carries on in the four dimensions of space-time.

Wiener believed that *negative feedback* explained goal-directed behavior in men *and* machines. Now, purpose is ordinarily taken as an *intentional* concept in the sense that some special relationship between a purposeful agent and the object of his actions is assumed in talking of goal-directed action; it is a relationship made special by its resolute inexpressibility in physical terms, unusual in that intentional acts need *not* be fastened on *existing* objects, as when Dr. Watanabe is said to pursue the Yeti.[38] Against this approach Wiener urged a fastidious behaviorism:

[35] Oskar Lange, *Introduction to Economic Cybernetics* (London: Pergamon Press, 1970), p. 24.

[36] On this score, see William T. Powers, "Feedback: Beyond Behaviorism," *Science*, Vol. 179 (January 26, 1973), pp. 351–356. Powers attempts to make good the deficiencies of behaviorism by adding to it a stock of concepts drawn from the theory of control. Now, making good the deficiencies of behaviorism is a strenuous project, and the concepts that Professor Powers brings to bear turn out to involve negative feedback merely. Powers, I gather, derives his work from an earlier joint effort, "A General Feedback Theory of Human Behavior," coauthored with R. K. Clark and R. L. MacFarland, to be found in *General Systems* (Washington, D.C.: Society for General Systems Research, 1960), pp. 63–83. The authors are correct, at least, in describing their theory as "general."

[37] See N. Wiener, A. Rosenblueth, and J. Bigelow, "Behavior, Purpose and Teleology," in Buckley, *Modern Systems Research*. Richard Taylor provided a rejoinder, Wiener and Rosenblueth a rejoinder to Taylor, and Taylor a rejoinder to their rejoinder, in turn. The whole kit and kaboodle can be found in Buckley.

[38] See R. M. Chisolm's entry, "Intentionality," in Paul Edwards, ed., *The Encyclopedia of Philosophy* (New York: Macmillan, 1967), for the details of this argument.

Given any object, relatively abstracted from its surroundings . . . the behavioristic approach consists in the examination of the output of the object and of the relations of the output to the input.[39]

And contrasted what he got with "functionalism"; *there*, questions of structure are admitted: analyzed are not only inputs and outputs but also the composition of the system putting through one to the other. In engineering terms, this resembles the distinction between input-output analysis and state-variable analysis. But Wiener quite clearly felt a strong *philosophical* commitment to behaviorism as a method conferring special virtues of scrupulousness and probity.

This is pretty sketchy. It was meant to be, I suppose, but Richard Taylor, in announcing that a "gross confusion" characterized cybernetics, had something worse in mind than simple sketchiness. Wiener, for example, wrote that

if the term purpose is to have any significance in science, it must be recognizable from the nature of the act, not from the study of or from any speculation on the structure and nature of the acting object.[40]

The trouble with this is that what ordinarily passes for purpose is *not* behavioristically definable: one act may embody a multitude of different purposes. In addressing an example offered by Wiener, Taylor argued that

the authors invite us to consider a car pursuing a man with the "clear purpose" of running him down. . . . *From observable behavior alone*, one cannot certainly determine what the purpose of the behaving object is, nor indeed whether it is purposeful at all. . . . Any driver who appeared to behave *as if* he were trying to run down a pedestrian, but who yet pleaded that he had no such intention, would not simply be *probably* lying but could not possibly be telling the truth.[41]

Professor Gerald Weinberg later remarked that "any undergraduate in computer science" could refute Taylor. But none of them has, and the argument between Wiener and Taylor threatens to become a permanent philosophical staple, with contrapuntal points reissued periodically. Some authors, including the meticulous Ernest Nagel, hew to a pound by pound increase in the complexity with which the definition of goal direction is phrased. By a goal-directed system Nagel means structures

that continue to manifest a certain property G . . . in the face of a rela-

---

[39] Wiener, Rosenblueth, and Bigelow, "Behavior, Purpose and Teleology," p. 221.
[40] Ibid.
[41] Richard Taylor, "Purposeful and Non-Purposeful Behavior: A Rejoinder," in Buckley, *Modern Systems Research*, p. 238.

tively extensive class of changes in their external environments or in some of their internal parts—changes which, if not compensated for by internal modification in the system, would result in the disappearance of G.

He follows this with pages and pages of cautious conditionals:

Let us now bring together these various points, and introduce some definitions. Assume S to be a system satisfying the following conditions: (1) S can be analyzed into a set of related parts or processes, a certain number of which (say three, namely A, B and C) are causally relevant to the occurrence in S of some property or mode of behavior G. At any time the state of S causally relevant to G can be specified by assigning values to a set of state variables 'A_x', 'B_y' and 'C_z'. The values of the state variables for any given time can be assigned independently of one another; but the possible values of each variable are restricted, in virtue of the nature of S, to certain classes of values K_a, K_b, and K_c, respectively. (2) If S is in a G-state at a given initial instant t_0 falling into some interval of time T, a change in any of the state variables will in general take S out of the G-state. Assume that a change is initiated in one of the state variables (say the parameter 'A'); and suppose that in fact the possible values of the parameter at time t_1 within the interval T but later than t_0 fall into a certain class K'_a, with the proviso that if this were the sole change in the state of S the system would be taken out of its G-state. Let us call this initiating change a "primary variation" in S. (3) However, the parts A, B, and C of S are so related that, when the primary variation in S occurs, the remaining parameters also vary, and in point of fact their values at time t_1 fall into certain classes K'_b and K'_c, respectively. These changes induced in B and C thus yield unique pairs of values for their parameters at time t_1, the pairs being elements of a class K'_{bc}. Were these latter changes the only ones in the initial G-state of S, unaccompanied by the indicated primary variation in S, the system would not be in a G-state at time t_1. (4) As a matter of fact, however, the elements of K'_a and K'_{bc} correspond to each other in a uniquely reciprocal manner, such that, when S is in a state specified by these corresponding values of the state variables, the system is in a G-state at time t_1. Let us call the changes in the state of S induced by the primary variation and represented by the pairs of values in K'_{bc} the "adaptive variations" of S with respect to the primary variation of S (i.e., with respect to possible values of the parameter 'A' in K'_a). Finally, when a system S satisfies all these assumptions for every pair of initial and subsequent instants in the interval T, the parts of S causally relevant to G will be said to be "directively organized during the interval of time T with respect to G"— or, more shortly, to be "directively organized," if the reference to G and T can be taken for granted.[42]

This represents an increase in explicitness when measured against the informal standards first set by Wiener, Rosenblueth, and Bigelow, but the underlying concepts have stayed pretty much the same. By goal

---

[42] Ernest Nagel, *The Structure of Science* (New York: Harcourt, Brace and World, 1961), p. 414.

direction, Nagel means an *observable* trait of *physically* organized systems; systems that are goal-directed maintain themselves against perturbations in compensatory fashion, as a yo-yo persists in periodic oscillations though its user has deflected the angle of inclination upward.

This process, of course, typically involves negative feedback; but the examples that actually fit suggest a certain porousness in the definition:

Now there are certainly many physico-chemical systems that are not ordinarily regarded as being "goal-directed" but that nonetheless appear to conform to the definition of directively organized systems proposed above,

Professor Nagel remarks, and he goes on to list such cases as pendulums at rest, elastic solids such as a sponge rubber ball, steady electrical currents flowing through a wire, and various thermodynamic equilibria. Nor is there any reason to stick with *thermodynamic* equilibria. Going down Nagel's list of conditions, it is hard indeed to see why every asymptotically stable system should not count as goal-directed once G has been identified as the equilibrium state itself.

Half the objections to Nagel's definition stress its insufficiency: some systems satisfy its strictures but are *not* goal-directed. Of no less oomph are objections that go the other way around: some systems *are* goal-directed or purposively organized but do not satisfy the definition. Against Wiener, Rosenblueth, and Bigelow, Taylor argued that "their conception automatically excludes certain types of distinctively and indisputably purposive behavior, by requiring that the goal be some *existing* object or feature in the environment of the behaving entity."[43] This makes it impossible to understand the behavior of an agent assiduously pursuing an object that simply does not exist. Nagel specifies no such goals, so the problem as Taylor frames it does not arise; still, G-states are meant to designate physically organized properties of an object, and precisely the same criticism may be made with such states in mind: in order to specify fully *which* G-state is animating a purposeful human organism mushing on toward the unattainable, one must have recourse to an essentially intentional vocabulary. Such states are *not* physically accessible: for an agent bent on devoting his life to the discovery of the largest prime, no accessible and purely physical G-state need do proxy for intentional states. The sheer physical behavior of an agent may manifest nothing remotely like goal direction: often it is only after his inner purpose has been discerned that we see in his haphazard, inconclusive, and easily perturbed gross physical behavior the outlines of directed activity.

[43] Taylor, "Purposeful and Non-Purposeful Behavior."

Input-output descriptions

More broadly, now, are there objections more philosophical to the whole notion that social systems instantiate somehow objects or processes best represented mathematically by devices taking inputs to outputs? It is this stress on the functional relationships between what happens to a system and what the system does when nudged, pricked, stimulated, set in motion, forced, or constrained that is distinctively system-theoretic.

To measure the merits of input-output analysis as a general schema, I hold constant scruples about the use of differential equations: the preliminary thesis, then, is in favor of some dynamical system such as

$$\frac{dx_i}{dt} = \sum_{j=1}^{n} a_{ij}(t)\, x_j + b_i(t), \qquad i = 1, 2, \ldots, n. \tag{3.37}$$

But mathematical systems theory begins not with (3.37) but with Figure 3.1. Some specific argument is needed to make the case that Figure 3.1 instead of (3.37) is what social systems need.

Now, (3.37) is inhomogeneous, and the play between homogeneous and inhomogeneous equations says something about the nature of inputs to a system. A system's *free response* is defined as the system's solution when the forcing functions $b_i(t)$ are set to zero. The solutions to homogeneous linear differential equations always appear as linear combinations of the linearly independent solutions:

$$y_{\text{free}}(t) = \sum_{i=1}^{n} c_i y_i(t). \tag{3.38}$$

A system's *forced response* involves solutions obtained when the initial conditions are set to zero: this highlights the input function alone. Forced solutions may be expressed by convolution integrals:

$$y_{\text{forced}}(t) = \int_0^t w(t - \tau) \left[\sum_{i=0}^{n} b_i(\tau)\, \frac{dx_i}{d\tau} \right] d\tau, \tag{3.39}$$

where $w(t - \tau)$ is the so-called weight function derived from the fundamental set of free solutions and the stock of constants c_i which vary with variations in the system's initial conditions under free conditions.

Thus the system's forced response is a function of the system's free response. This makes the point that general solutions to inhomogeneous linear equations with variable coefficients are derived via linear algebra from corresponding homogeneous equations. The *total response* of the system is defined simply as the sum of forced and free responses.

Inputs form one part of the conceptual burden of mathematical systems theory; *outputs*, the other. In the classical tradition outputs arise as solutions to differential equations. An example is the circuit equation

$$\frac{dy}{dt} + \frac{y}{RC} = \frac{u(t)}{RC},$$ (3.40)

where R is resistance, C is capacitance, and $u(t)$ and $y(t)$ are input and output voltages, respectively. To specify $y(t)$ uniquely, of course, some initial conditions must be added.

It may happen that solutions to a system are inaccessible: the possibility is suppressed on the classical view inasmuch as outputs are *identified* as solutions. But even when the dynamics of the plant are expressed in terms of state variables satisfying some vector-matrix differential equation, the system's output can usually be described as a linear function of the solution set to the system's differential equations. This makes for a one-step elaboration in the concept of an output.

Classical dynamical systems are quite often understood as free systems of linear ordinary differential equations with constant coefficients. So long as dynamical structure is what is needed, classical theory has no use for inputs and outputs, and for good reason. If the underlying system is inhomogeneous and linear, solutions may be analyzed as a superposition of the linearly independent solutions to homogeneous equations, so that the solutions to the free system are still the only ones that count. Having the free system, one has the system's outputs; having the inhomogeneous system, one has the inputs as well. The differential structure of the system trivially generates both inputs and outputs, and an analysis basically dynamical in character that yet reconstrues the plain dynamical facts to conform with Figure 3.1 represents *systemic inflation*.

Professor Milsum provides an example.[44] The simple equation governing exponential growth he displays on p. 21 of his text:

$$\frac{dx}{dt} = kx.$$ (3.41)

---

[44] Milsum, *Biological Control Systems Analysis*, pp. 21–22.

The conceptual innocent would insist that here there is nothing save
the trivial or zero input, and not much past the trivial output either
since this is a simple first-order linear equation with a constant coefficient. To be sure, Milsum sees nothing more in (3.41). Nonetheless, by
p. 22 a conversion of considerable grandeur turns up, and a diagram
with miraculously many arrows and nodes comes to extinguish one's
instinctive sense that in (3.41) one has neither inputs nor outputs but
a differential expression merely.

Am I urging, then, that any use of Figure 3.1 as a schema for the
social sciences constitutes a conceptual solecism? Not at all: I mean to
indict only large-hearted appeals to Figure 3.1 under conditions where
the appropriate differential equations would have done as well.

Control theory

These conditions are not general. In the theory of linear systems and in
classical or modern control theory, the chief studies no longer involve
methods merely for the representation of the plant's dynamical properties as a dynamical system. Professor Brockett, for example, describes
a linear *system* as a pair of equations,

$$\frac{dx(t)}{dt} = A(t)\,x(t) + B(t)\,u(t),$$

$$y(t) \quad = C(t)\,x(t) + D(t)\,u(t),$$

(3.42)

and observes that such an object is

more complicated than just a linear differential equation and there are
many interesting questions which arise in the study of linear systems
which have no counterpart in classical linear differential equation theory.

What *is* of interest is questions

which relate to the dependence of y on u which these equations imply.
The point of view one adopts is that u is something that can be manipulated and that y is a consequence of this manipulation.[45]

It is this inflexible stress on the *relationship* between inputs and outputs that makes systems theory something more than a baroque way
of expressing obvious differential thoughts: it is what one gets in Figure
3.4 when the full control-theoretic apparatus supplants Figure 3.1. Nor
is control theory in all its fullness Figure 3.1 enriched by systems of

[45] R. W. Brockett, *Finite Dimensional Linear Systems* (New York: John Wiley &
Sons, 1970), p. 67.

simultaneous equations exhibiting feedback. The Volterra-Lotka equations of Chapter 2 show simultaneous and reciprocal influence—feedback but not control. Crucial to control systems are the concepts of a target output and an error signal, and these are not elements of simple simultaneous systems.

The concerns of control theory have historically arisen under circumstances demanding regulation. This naturally leads to the theory of *design*, in which premiums are placed on techniques for compensation, control, regulation, modulation, and modification. Heirs to these concerns are pursuits such as *optimal control theory* in which controls minimizing certain functionals are sought. Optimal control theory has had some successes in economics and in game theory; its triumphs provide some justification for methods that characterize a system as a device taking inputs to outputs.

There is room also for what might be called the *analytic* theory of control systems: if the premises of optimal control theory demand that the perfect path be hunted out, the assumptions of analytic control theory say simply that society embodies specific control mechanisms: it is the analyst's task to uproot and then describe them. Of necessity, this will involve more than a description of their dynamic characteristics—more, too, than might be expressed merely by differential equations. Outside of economics and the more specialized traditions of optimal control theory, I know of little work in the social sciences being pursued strictly as analytic control theory.

Speculative biology is rich in such applications, however, and a good example is provided by the Pennock-Attinger model of the oxygen transport system.[46] These authors content themselves with description and analysis: the various parts of the oxygen transport system are laid out, mechanisms of control are identified, state equations are introduced, and the relationships between the oxygen demands of the heart, the tissues, and the respiratory organs are sketched. Or there are the sparse examples that Professor Milsum provides: in the course of a chapter on the special problems of biological control systems, a rather elaborate chart depicting the "spatial configuration of the visual fixation system" is laid out; later there are equations. A sober air of hypothesis pervades throughout. There is nothing really of optimal control theory in all this since both Milsum and Pennock and Attinger mean to describe the systems that they find—and these may well not be optimum. But classical and modern control theory play a necessary role

[46] B. Pennock and E. O. Attinger, "Optimization of the Oxygen Transport System," *Biophysics Journal*, Vol. 8 (1968), pp. 879-896.

here: to understand a fully control-theoretic object one requires a control-theoretic language.

A social scientist studies social systems to get at their laws. This leaves open the possibility that his studies will coincide with the engineer's: systems analysts argue that *their* grand aim is to uncover the regular and lawlike structure animating objects such as those described by Figure 3.4. This tends to collapse the distinctions between engineering and social sciences; it undercuts, too, such plausible points of distinction as are listed by Professor Grodins:

> The engineer specifies the nature of "desirable" or "optimal" behavior, and he knows the anatomy and physiology of the system because he built it. On the contrary the groping biologist is confronted with an existing system designed and built for a mysterious purpose by an unknown engineer. . . . Thus whereas the engineer primarily deals with the direct and design problem, the biologist is faced with the inductive and inverse problem.[47]

This is to play up excessively the quantitative degree of groping that characterizes the working biologist: given a known but complex system, the engineer can respond with assurances that he, too, gropes considerably in order to come up with theorems that will enable him to get on with the business of design. Both optimal control theory and what I have called the analytic theory of control seem much of a piece with social sciences generally.

If there is no intrinsic distinction between the aims of the engineer and those of the social scientist, and hence no criticism founded on a confusion of roles, there is a distinction in the *sequences* of the sciences. In the classical tradition, there is no engineering without some prior specification of a dynamical model; Professor Kalman writes that "the model is obtained from physical measurement or physical laws; its actual determination is outside the scope of control theory or even systems theory."[48] Kalman quite properly thinks of the determination of physical laws as ancillary. If the models fit the definitions of control theory, draw out the theorems and apply them blindly to whatever situations the dynamic equations describe. *That the models meet the criteria of control theory, or systems engineering generally, is an assumption and not a consequence of the systems-analytic technique.* And it is an assumption of some strength; in the classical tradition, at any rate, the models are severely specialized: differential equations

---

[47] Grodins, "Similarities and Differences Between Control Systems in Engineering and Biology," p. 4.
[48] Kalman, Falb, and Arbib, *Topics in Mathematical System Theory*, p. 26.

first, single input second, constant coefficients third, linearity fourth. The social scientist prepared to make the case for control theory, or even for the vastly more flabby thesis that systems are somehow input-output devices, must *logically* stand ready with a suitable dynamical theory: there is no looking to control theory itself for such a structure.

Thus the imagery of control systems in the end offers a double rejection: the models that it demands are inherently not very much like social or biological or political control systems, and, what is worse, it offers no prescription at all for finding the right systems but requires instead that they be available if the application of control theory is to make any sense at all.

Automata Theory
and Molecular Biology

Ce n'est pas que les assimilations de la mécanique vitale à certain aspects de la technique humaine (automates, ordinateurs électroniques, etc.) soient sans valeur; mais ces comparaisons ne peuvent jouer que pour des mécanismes partiels, tout montés, et en pleine activité fonctionnelle; elles ne sauraient en aucun cas s'appliquer à la structure globale des êtres vivants, à leur épigénèse et leur maturation physiologique.

René Thom[49]

Mathematical systems theorists, ever omnivorous, consider virtually anything that fits Figure 3.1, however incongruously, to be within their province. The classical tradition requires that the flow of time be·represented by the real numbers. This makes possible the application of the entire body of continuous mathematics. But time may also be taken as *granular*, with intervals fixed by integers. *Automata theory* is one discipline that then results; its systems are quite different from the usual run of rheostats, filters, transducers, and amplifiers in that *discrete machines* move inputs to outputs. The modern theory of automata covers much ground.[50] There is the abstract theory of the machine, of course: in *Turing machines* automata theory makes contact with the theory of recursive functions. There is also the theory of nerve networks and the theory of self-reproducing automata, various aspects of the theory of design, control, and regulation, theories of pattern recognition and grammar, and theories of computability and cost. An exquisitely combinatorial air pervades throughout. That such devices fall under the aegis of Figure 3.1 is evidence that in mathematical systems theory there is independent origin and convergent evolution, much as the narwhal and the porpoise both discovered the external bladder.

[49] René Thom, *Stabilité Structurelle et Morphogénèse* (New York: W. A. Benjamin, 1972), p. 207.
[50] On automata theory, see R. J. Nelson, *Automata Theory* (New York: John Wiley & Sons, 1968); for self-reproducing automata, see von Neumann's posthumous monograph, *Theory of Self-Reproducing Automata*, edited by A. W. Burks (Urbana, Ill.: University of Illinois Press, 1966). For nerve nets, see Burks and Wright, "Theory of Logical Nets," *Proceedings of the IRE*, Vol. 41, pp. 1357–1365. For grammars, see Noam Chomsky, "Formal Properties of Grammars," in R. D. Luce, R. R. Bush, and E. Galanter, eds., *Handbook of Mathematical Psychology*, Vol. 2 (New York: John Wiley & Sons, 1963).

Automata theory and the classical theory of linear systems thus offer rivalrous interpretations of systemhood and its essences; in so doing the theories tap different intuitions. The classical tradition abstracts from common engineering experiences, stripping from filters, amplifiers, and electrical circuits their coarse physical properties and exhibiting what remains in the sparse but satisfying language of mathematical analysis and the theory of ordinary differential equations. Automata theory, on the other hand, has historically arisen (at least in part) as the result of abstractions performed on *computational* experiences. The digital computer is not so much the object *from* which automata theory has performed successive abstractions as the physical expression of automata-theoretic principles made palpable only after they have been derived from contemplations on the nature of computation itself.

Systems theorists have pursued automata for much the same reasons that they have chased hungrily after the engineers[51]—to get hold of general and integrative concepts broad enough to explain phenomena that at some inaccessible level obviously display properties of cohesion, organization, control, and regulation. Social and political scientists have had a marked preference for the fluid methods of the classical tradition. Speculative biologists have made use of such methods, but there is also a strong and interesting tradition, beginning perhaps with the work of McCulloch and Pitts and of von Neumann, dedicated to automata-theoretic models for basic biological processes—the growth of intelligence, the connection of neurons, the development of the brain, the phenomenon of self-replication, the mastery of grammar, and the divisions of the cell. Theorists such as von Neumann or Noam Chomsky adopted automata-theoretic tools to explore the outer limits of *algorithmic accessibility*. Machines that reproduce and function reliably despite unreliable components, or that generate the sentences of a natural language, provide the skeptic with something like an existence proof: the ebb and flow of the generations and the creation of language represent nothing *beyond* our algorithmic experience.[52]

Finite-State Machines

Automata theory, like the theory of linear systems, takes certain objects as fundamental. The Turing machine, described briefly in Chapter 1, embodies the *machine* taken to the outermost limits of idealization;

---

[51] See M. Arbib, "Automata Theory and Development," Part I, *Journal of Theoretical Biology* Vol. 14 (1967), pp. 131–156.
[52] See Chomsky's "Formal Properties of Grammars" for details.

here is a concept of computability framed without regard to time, power, speed of computation, or ease of access. This is plainly the area in which automata theory achieves its deepest results: since Turing machines generate all and only the recursively enumerable sets, the theory of Turing machines is an alternate method of expressing the theorems of recursive function theory. But Turing machines are infinite: they have unbounded memories and their characteristic mode of behavior is in many respects elusive. Given a set of instructions, the Turing machine will complete any given sequence in a finite amount of time, but if the sequence is unperformable, the machine may grind on forever. Considering the instructions to a given Turing machine, there is no effective way of determining whether the machine will be able to execute the task at all; if it can execute assigned sequences, there is no way of telling how long it will take, how much tape will be required, or whether the completed output will be finite or infinite.

So Turing machines are interesting only as objects that reflect certain characteristic intellectual powers and properties; in this sense, their usefulness goes quite beyond the area of the theory of recursive functions since human capacities apparently require representation in infinitary terms. Thus it falls within ordinary human competence to multiply pairs of numbers quite without upper bound, and only a device with an unbounded memory could mirror this capacity. Nonetheless, normal intellectual activities are subject to the inescapable limitations of time and memory, and *this* suggests a device with finite memory. It is just this double image that has suggested to Chomsky the distinction between *competence* and *performance*, with intrinsic cognitive abilities modeled by infinite devices, and performative capacities by finite devices.

Turing machines involve powerful idealizations; finite automata, manageable memories and mathematics. The distinction between competence and performance may well turn up in the analysis of biological phenomena; but so long as automata-theoretic models of processes actually instantiated by objects such as the bacterial cell, the cerebrum, or the central nervous system are at issue, strongly idealized automata are conceptually luxurious.

This is the motivation for taking as central such devices as the finite-state automaton: when I speak of centrality, I mean merely that the finite-state automaton, like the simple linear system, is the basic intellectual object of the theory and the model against which claims for mathematical usefulness must be tested.

Essentially, a finite-state machine may be depicted as a reading head

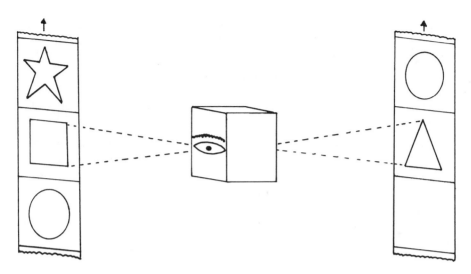

Figure 3.10
A finite-state machine.

mounted between two tapes (Figure 3.10).[53] The tapes are subdivided into squares, and symbols appear on the squares of the input tape. Surveying a given symbol at time t, the reading head consults its internal instructions and then moves to print an appropriate symbol on its output tape. That having been done, the machine clicks to time $t + 1$, looks to a new square on its input tape, and repeats the operation.

This informal description should call to mind the Turing machine. Were the tapes infinitely extendable, finite-state machines would be Turing machines. But the description should also evoke the input-output device of Figure 3.1. Finally, finite-state machines should suggest computer programs since both work via the sequential execution of fixed steps.

In more detail, then, a finite-state machine may be identified with a 5-tuple $\{A, S, Z, \nu, \zeta\}$, where $A = \{a_0, a_1, \ldots, a_n\}$ is a finite list of *input symbols*; $Z = \{z_0, z_1, \ldots, z_n\}$, a finite list of *output symbols*; $S = \{s_0, s_1, \ldots, s_r\}$, a set of *internal states*; ν, a *next-state function* from $S \times A$ to S; and ζ, an *output function* from $S \times A$ to Z. Birkhoff and Bartee remark that

a finite-state machine is therefore defined mathematically by three sets

---

[53] Here I follow Garrett Birkhoff and Thomas C. Bartee, *Modern Applied Algebra* (New York: McGraw-Hill, 1970).

and two functions. It acts by "reading" or "accepting" a sequence or "string" of input symbols (the *program*) and then "printing" or delivering output symbols. A finite-state machine starts in some internal state s_j; it then reads the first input symbol a_k. The combination of s_j and a_k cause an output symbol z_1 to be produced through the function ζ and the machine to go to a new state s_r as directed by the function ν. After this, the machine reads the next input symbol, and so on until it finishes the string.

A very simple example is afforded by the following machine. M has a two-symbol input alphabet $A = \{0, 1\}$, a two-character output alphabet $Z = \{0, 1\}$, and three internal states $S = \{s_0, s_1, s_2\}$. Next-state and output functions are the following:

$\nu : (0, s_0) \mapsto s_1$ $\qquad$ $\zeta : (0, s_0) \mapsto 0$
$ (1, s_0) \mapsto s_0$ $\qquad (1, s_0) \mapsto 1$
$ (0, s_1) \mapsto s_2$ $\qquad (0, s_1) \mapsto 1$
$ (1, s_1) \mapsto s_1$ $\qquad (1, s_1) \mapsto 0$
$ (0, s_2) \mapsto s_0$ $\qquad (0, s_2) \mapsto 1$
$ (1, s_2) \mapsto s_2$ $\qquad (1, s_2) \mapsto 0$

This is a machine that will, for example, take the input tape

$0 : 1 : 0 : 1 : \ldots$

to the output tape

$0 : 0 : 1 : 0 : \ldots .$

Given that the machine starts in state s_0, its next move, as it scans 0 on the input tape, is to print 0—this by the first instruction for ζ; having printed 0 it moves to state s_1—this by the first instruction for ν. But this means that the machine is now in state s_1 as it scans 1 on the input

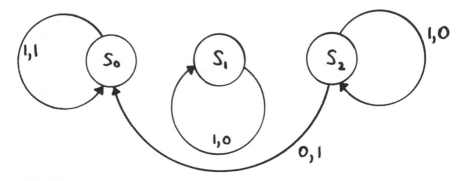

Figure 3.11
State diagram for the finite-state machine discussed in the text.

tape. This combination fixes its next move and its next printout; and so on until the output tape is completed.

Such operations may also be represented by *state diagrams*: thus, for the machine just given, one has Figure 3.11. Here nodes correspond to states and arrows to transitions determined jointly by inputs and outputs. The leftmost arrow, for example, indicates that the machine, starting in state s_0 and scanning 1, prints 1 and remains in the same state.

Molecular Biological Systems

The analogies between a living organism and a machine hold true to a remarkable extent at all levels at which it is investigated.

Jean Pierre Changeux[54]

It is now almost twenty years since Watson and Crick discovered that deoxyribonucleic acid (DNA) is formed of a double helix held together lightly in midcell by pairs of hydrogen-bonded bases. The astonishing spasm of creative work that began in 1951 or so is now ending; some molecular biologists are deserting their field entirely, convinced that the essential work has been done and that only the tedious business of making plain the details remains. Watson strikes just the right note of hard-won contentment when he writes that

complete certainty now exists among essentially all biochemists that the . . . characteristics of living organisms . . . will all be understood in terms of the coordinative interactions of small and large molecules. Much is already known . . . , enough to give us confidence that further research . . . will eventually provide man with the ability to describe with completeness the essential features that constitute life.[55]

Such briefs for the powers of biochemistry have been common enough: ever since Wohler synthesized urea from lead cyanate and ammonia, biologists have been pressing for an absorption of their subject into physics or chemistry. But only with the extension of biology to the molecular level has this claim come to seem other than the merest braggadocio. Professor Lehninger's magnificent text in biochemistry now commences confidently with the assertion that biology simply *is* chemistry, although of an especially intricate and peculiarly combinatorial kind.[56]

[54] Jean Pierre Changeux, *Scientific American* (April 1965), p. 36.
[55] James D. Watson, *The Molecular Biology of the Gene* (New York: W. A. Benjamin, 1970), p. 67.
[56] Albert L. Lehninger, *Biochemistry* (New York: Worth, 1970), pp. 3-13. Watson and Lehninger between them will give the biologically inclined a sense of the pres-

Of course, both Lehninger and Watson would consider an explanation uncouched in physical or chemical terms incomplete; at the same time they leave the notion of a *chemical* explanation in a state of perfect obscurity—no doubt to accommodate as chemical virtually any statement about the biological molecules. Those details, whose clarification was to be a matter of patient but entirely routine analysis, can bulk large. The greater part of work to date has been directed toward the bacterium *Escherichia coli*. Here biologists are dealing with a system that contains some five thousand organic compounds, including three thousand or so proteins and over one thousand nucleic acids. Many are of a complexity so outrageous that no one expects their eventual unravelment. And only one-third to one-fifth of *E. coli*'s metabolic reactions are known. A second generation of molecular biologists is just beginning to extend phage and bacterial techniques to mammalian systems. Possibly these systems will prove to be as the elephant is to the mouse: much the same, only larger. Some biologists apparently believe this. Others, of a gloomier but probably more realistic cast of mind, expect that higher organisms will exhibit a *qualitatively* greater order of complexity than bacteria. A human being, for example, contains over five million separate proteins, none of them identical with any found in *E. coli*. A mysterious process of differentiation takes place shortly after fertilization, as one cell gives rise to an ensemble of cellular empires arranged in tissues capable of the most exquisite biochemical activity. There are obscure processes of control and information storage; information is passed between cells and organized by cognitive mechanisms whose biochemical properties are largely unknown.[57]

All this might give rise to a sense of discouragement about the larger prospects for molecular biology; but the work on *E. coli* has given biochemists a comforting feel for the sort of general schema that must underlie the endless unconquered detail available in systems more complex than bacteria.

Periodically, almost everything alive contracts to the point of a single cell and then reassembles itself into a faithfully drawn, full-fledged organism. Generations vary, but by and large the endless cycles of contraction and expansion move along an astonishing amount of biological information. The sheer quantity and quality of this achievement is

---

ent state of molecular biology. Zealots will want to consult the volumes of the *Cold Spring Harbor Symposium on Quantitative Biology* for original papers and a mass of stultifying detail.

[57] To speak of the molecular biology of memory or learning is still to speak of a perfect vacuum. In this regard, see George Ungar, ed., *Molecular Mechanisms in Memory and Learning* (New York: Plenum Press, 1970).

surely unique in the physical world. No less is true of single-celled organisms: *E. coli* or cells within a complex mammal.

The fact that in reproduction life contracts to a single cell suggests that the cell contains the material for a later orchestration of itself into a complete organism. It is conceivable, of course, that the mechanism is purely physical, with each organism lugging about a miniature version of its own posterity. Much more likely, however, even on a priori grounds of reasonableness, elegance, and simplicity, is the hypothesis that the cell contains what Erwin Schrödinger called a *code script*, a sort of cellular amanuensis, set to record the gross and microscopic features of the parental cell and pass the information thus obtained to the cell's descendants.[58]

Such is the idea one might reach by considering reproduction as a problem in pure thought. An approach so fashioned should not be scorned on account of its abstractness: saving biology from the sorrows of biochemistry was in large part a theoretical enterprise initiated by physicists in the mid to late 1940s. The code script that Schrödinger anticipated turns out to be DNA, a very large macromolecule composed of linked mononucleotides built up from a purine or pyrimidine base, one molecule of 2-deoxy-D-ribose, and one molecule of phosphoric acid. The nucleotides differ only with respect to their bases: adenine, guanine, thymine, and cytosine (A, C, T, G). Within the cell, DNA usually appears as a double helix, with one strand matched to the other through complementary pairing of the bases: A with T, C with G. Hydrogen bonds provide the molecular glue holding the bases together.

The curiously elegant configuration favored by the nucleic acids suggests how a cell that splits in two manages to partition its code script between its offspring. As cell division takes place, the DNA molecules rotate about an axis; since the hydrogen bonds between base pairs are weak, the strands begin to unravel; as they do, an enzyme—DNA polymerase—catalyzes a recombination reaction between the strands and four nucleotide precursors. Each separate strand then reassembles into a stable double-stranded helix as a result of the intermolecular forces binding the base pairs of nucleotides. The energy for this reaction derives from the hydrolysis of pyrophosphate, formed when the DNA chain is forged. The liberation of some 14,000 calories gives the reaction considerable thermodynamic thrust.

It is the alternation of the four bases along the strands of DNA that allows the molecule to store information. This is not a sparse system. A molecule of DNA having a molecular weight of 1,000,000 will con-

---

[58] See Erwin Schrödinger, *What Is Life?* (New York: Macmillan, 1945).

tain a string of some 4000 nucleotides. The specification of any given base requires $\log_2 4$, or 2, bits of information, so the informational content of the strand as a whole comes to 8000 bits. Analogical reasoning suggests that a simple bacterial cell contains about 10^{12} bits of information. Some 4^{1500} messages can be expressed by nucleotide sequences—another number of reassuringly large size.

The bacterial cell, of course, is something more than a clutch of nucleic acids. Proteins, which are globular macromolecules constructed from a stock of some 20 amino acids, displace most the the cell's organic space; carbohydrates and lipids make up the balance. Typically, a completed protein forms a chain of 250 amino acid residues held together in an assembled molecule by dipeptide bonds. In the possibilities for order among the residues lie all the differences between a man and a mouse. So the sense of genetic identity that marks the bacterial cell *as* a bacterial cell must somehow be stamped into sequences of amino acids. The information for that achievement is contained in the nucleotides, expressed in the proteins.

The four nucleotides, we now know, group themselves into a triplet code of 64 codons. A string of nucleotides is thus composed of alternating three-letter words. The amino acids are matched to the codons, with some residues multiply coded and some codons reserved as marks of punctuation. In the transcription and translation of genetic information, the linear order of the nucleotides serves to induce a corresponding linear order first onto messenger RNA (mRNA) and then onto the amino acids. The sequential arrangement of the amino acids in turn fixes the ultimate stereochemical configuration of the completed protein: out of materials no more impressive than the capacity to rank matched codons and amino acids in parallel arrays, the bacterial cell manages to operate a channel of communication of almost unlimited scope and complexity.

Biologists often allude to the steps so described as the central dogma and trot out the following diagram:

$$\text{DNA duplication} \xrightarrow{\text{Transcription}} \text{RNA} \xrightarrow{\text{Translation}} \text{Protein.}$$

It is a dogma and a diagram of some force. Enclose the diagram in a circle representing the cell wall and you have a remarkable schema of the very most basic processes involved in the living state. Protein structure results from molecular translation, and the *regulation* of the bacterial cell boils down to the business of sequencing the right protein at the right time. The *E. coli* enzyme β-galactosidase, to take an example, splits lactose into glucose and galactose. Breakdown rates vary in proportion to the quantity of β-galactosidase within the cell; this, in turn, is proportional to the amount of lactose in the growth medium. Lactose is an *inducer*, and β-galactosidase is an *inducible enzyme*. The bacterial cell also contains repressors that act to inhibit the transcription of β-galactosidase. Repressors are themselves proteins, coded by the so-called regulatory genes: in suppressing transcription they bind to DNA sites called *operators*. When the cell grows on a lactose-rich medium, repression is itself repressed: the induction of β-galactosidase proceeds by rendering the corresponding repressor inactive. The elimination of lactose from the growth medium frees the repressor, and the transcription of the enzyme is slowed.[59]

The regulatory mechanisms of the LAC system are of compelling intricacy. No doubt, biologists will mark my description as baldly truncated. Still, one can see how the fine tuning falls within reach of the central dogma: regulation is entirely indirect, and involves the translation—from DNA-stored information—of superbly conceived and rather specialized regulatory proteins.

Information, of course, is not passed from one generation to another with perfect fidelity; mutations occur. But from this trifling observation, the central dogma bends to an explanation of *speciation*. There is currently some debate between those who believe that the mutation rate is high and hold to a doctrine of selective neutrality and those who hew to the more standard versions of neo-Darwinism.[60] Seen from a sufficient distance, the controversy collapses into a quibble, with both sides agreeing basically that biological change and the whole gigantic panorama of life in its various forms are a matter merely of a system that misfires randomly if regularly and then manages to trap its usable mistakes. So the enlarged central dogma, or paradigm (in the sense of Thomas Kuhn), encompasses an account of the cell's ability to store, express, replicate, and change information. These are the fundamental features of life, no less, and a schema that says something interesting about them all has at least scope to commend it.

---

[59] See F. Jacob and J. Monod, "Genetic Regulatory Mechanisms in the Synthesis of Proteins," *Journal of Molecular Biology*, Vol. 3 (1961), p. 318, for the details. Monod has written a popular account of his philosophical convictions in *Chance and Necessity* (New York: A. A. Knopf, 1971).

[60] See M. Kimura and T. Ohta, *Theoretical Aspects of Population Genetics* (Princeton, N.J.: Princeton University Press, 1971).

Schemata for molecular biological systems

Critics have on occasion surveyed the substance of molecular biology in a spirit of less than perfect appreciation. Eugene Wigner has written of the impossibility of self-replicating mechanisms—this on quantum-mechanical grounds; and Georg Kreisel has seen in the apparently recursive functions of subcellular biology a system that could not possibly account for unrecursive behavior, among mathematicians, say.[61] Biologists, too, have had doubts: Macfarlane Burnett has argued that the mysteries of mechanisms such as in the immunodefense system may simply be too complex to track down to the molecular level.[62]

Philosophers have generally remained undebauched by anything like a precise acquaintance with modern biochemistry; still, they might look to the last paragraphs of Wittgenstein's *Investigations* for a sense of the spirit that they will require. There Wittgenstein mentions experimental methods and conceptual confusion, almost as if the two were inevitably yoked together.

Concepts such as code, information, language, control regulation, and translation come easy to the biologist as he describes the cell according to lights provided by the central dogma. These are distinctly theoretical notions, and some researchers have the laudable ambition of setting biological entities in a context amenable to abstract description and analysis. Such settings would provide *formal models* for the bacterial cell: given some interesting theory, one might hope to discover that *E. coli*, under suitable and no doubt very powerful idealizations, was in at least some respects isomorphic to its models. But it is no easy matter to hit on the right theories—this despite the naturalness of many biological concepts.

The thought that DNA is in fact a *code* might drive the biologist toward a celebration of the powers of formal code theory. The algebra of semigroups comes naturally into play here: whereas a group is a set closed under an associative binary operation, containing both inverse and identity elements, a semigroup (or monoid) has no inverse. The set of all finite sequences drawn from an arbitrary alphabet becomes a monoid on the assumption that the binary operation is associative. Given two monoids, with alphabets Σ and Υ, a code can be defined as

---

[61] Eugene Wigner, "The Probability of the Existence of a Self-replicating Unit," in *The Logic of Personal Knowledge: Essays Presented to Michael Polanyi on His Seventieth Birthday* (Glencoe, Ill.: The Free Press, 1961); Georg Kreisel, "Mathematical Logic," in R. Schoenman, ed., *Bertrand Russell, Philosopher of the Century* (Boston: Little, Brown, 1967).

[62] MacFarlane Burnett, *Immunological Surveillance* (New York: Pergamon Press, 1970), pp. 236–244.

an injection I of strings in Σ into strings in Υ such that if $\nu_i, \nu_j \in \Sigma$, then $I(\widehat{\nu_i\nu_j}) = \widehat{I(\nu_i)I(\nu_j)}$. I is thus an isomorphism between strings in Σ and a subset of the strings in Υ; such "spelling isomorphisms" provide a complete device for shuttling between the monoids whose alphabets are set at Σ and Υ.[63]

Many different codes can be described within this framework: tree codes, self-synchronizing codes, anagrammatic codes, and so forth. There is thus some temptation to see DNA as a formal code-theoretic object. The four nucleotides might form the elements of a base vocabulary X; the 4^3 codons formed from the elements of X might form an extended vocabulary X^*. The DNA monoid G is then defined as the set of all sequences composed of elements drawn from X^*, including sequences of length 1. Concatenation is the group-theoretic operation, and the null sequence functions as the identity element. The transmission of information first from G to mRNA and then to the amino acids is definable in code-theoretic terms, with X^* playing the role of Σ and strings of amino acids drawn from a base vocabulary of twenty residues the role of Υ.

The early history of work on the nucleic acids was marked by an experimentally unsullied appreciation for abstract methods of this sort. The journals were heavy with dense articles on the coding problem; for a time the fact that there are just twenty amino acids became a subject occasioning numerological awe. The discovery in 1961 that polyuracil codes for polyphenylalanine brought this line of speculation to an end: since then biochemists have taken pleasure in the thought that decisive progress in the unraveling of the code was achieved in a way that owed nothing to mathematical methods.

Still there are things that can be learned from code theory. George Pieczenik has argued that the nucleic acids embody a *self-synchronizing* code: codons that usually mark the end to a string of nucleotides also crop up just out of phase at the beginning of many sequences, where they are able to bring a misread message to a halt.[64] This, I think, is a striking result which hints at the existence of some fairly elaborate genetic structures. But code theory does comprise an object of fragile powers. An instructive example is afforded by the Morse code, a device that effects an interchange between a vocabulary of dots and dashes and the full stock of letters of the English alphabet. From a code-theo-

---

[63] See the useful discussion of codes in N. Chomsky and G. A. Miller, "Introduction to the Formal Analysis of Natural Languages," in Luce, Bush, and Galanter, *Handbook of Mathematical Psychology*, Vol. 2.

[64] *Proceedings of the 2nd Graduate Research Conference in Genetics and Molecular Biology*, June 12-14, 1970, Wesleyan University, Middletown, Conn.

retic point of view, *any* string of dots and dashes is suitable for such exchange; but in using the code one takes for granted some prior sense of sentencehood, so that only strings carrying meaning are actually subject to coding. The algebraic properties of the code are of little use here: the Morse code easily arranges for an exchange of bilateral nonsense. Presumably the same thing is true of the genetic code. In this respect, code theory says nothing about the *organization* that may be in evidence in the nucleic acids: every string of codons comprises a codable object. This is hardly a defect in a theory created to deal with problems that arise when information is passed along a channel of communication. But a sense of algebraic rapture shifts the biologist's attention to the medium much at the expense of the message; deflated, he will require concepts that code theory cannot express.

Information theory has found some favor because there is a useful distinction between the informational macromolecules—the nucleotides and the proteins—and the other cellular constituents.[65] But I suspect that *automata theory* will come to seem the source for the most natural models of the bacterial cell. The theory of machines embraces a purely theoretical attitude toward computation and so is properly a branch of recursive-function theory. An air of unconcern for the dismaying realities of computational cost, hardware, efficiencies of scale, real time, and the like give the theory some breadth: automata-theoretic tactics are deployed in discussions of neural nets, grammars, switching circuits, and a minor batch of disciplines that seem as much united by their adoption of automata as by anything else. In its most general form, automata theory treats of the conversion of *inputs* to *outputs* via a device that admits of *states*. The conversion is effected between items that are discrete and combinatorial: numerals, words, sentences, letters. The bacterial cell seems distinctly amenable to this sort of treatment: sequences of nucleotides resemble sentences; codons, words. The very intricate processes of control and organization evoke computational machinery of various kinds.

Other researchers have hit on this idea, especially those who have read von Neumann's work on self-replicating automata.[66] Biologists have a natural interest in exploiting automata-theoretic concepts in areas beyond the reach of classical biochemical techniques; thus they

[65] For the information-theoretic point of view, see Lila L. Gatlin, *Information Theory and the Living System* (New York: Columbia University Press, 1972).
[66] See, for example, Aristid Lindenmayer, "Mathematical Models for Cellular Interaction in Development," *Journal of Theoretical Biology*, Vol. 18 (1968), pp. 280–299; W. Stahl and H. Goheen, "Molecular Algorithms," *Journal of Theoretical Biology*, Vol. 5 (1963), pp. 266–287.

have assayed models of such mysteries as cellular differentiation and development. The idealized cell appears as an automaton of a certain kind and complexity in this sort of research, and the associated programs are designed to push the machine through maneuvers that look vaguely biological. Often the machines turn out to have very strong computational powers. Still, the thought that the cell is a machine has its attractions: it invokes a familiar class of objects—cellular models— while apparently vindicating a very old philosophical perspective. Allures on this order have even brought Jacques Monod to the confession that it is just in the discovery of the machinelike nature of the cell that modern biology has had its most impressive triumphs.[67]

Philosophically, one would like to see the analogy between bacterial cells and automata pursued in a more general setting, without fixed commitments to specific devices. Turing machines are able to generate all the recursively enumerable sets, so models for the cell can be no *more* powerful than Turing machines if the cell does anything at all that can reasonably be called computation. There is no special reason to assume the bacterial cell capable of such computational profligacy, however, so the most natural automata-theoretic models will probably be drawn from the class of finite-state machines.

Although such machines have a certain overall simplicity that makes them especially attractive as models for the bacterial cell, the fact that they incorporate no provisions for storing information is a decided liability. This can be made good by the addition of storage or "stack" capacity. *Push-down-storage automata* (PDSA) are finite-state machines enriched with such capacity. Storage is accomplished on a first-in/last-out basis, but the stack itself can be read in either direction and is available both for reading and for recording symbols. PDSA are usually defined as machines that accept or reject input strings: a given string is accepted if the machine reads it when the stack is empty. But the difference between generators and acceptors is more or less terminological: the equivalence between the full class of PDSA and the context-free generative grammars is well known. A PDSA, then, can be defined as a 6-tuple $< X, S, \Delta, \delta, \epsilon, s_0 >$, where X is a finite set of input symbols, S is a finite set of input states, $\Delta \supset X$ is the set of input symbols plus a set of auxiliary symbols for entry on the stack, δ is a function from $X \times S \times \Delta$ to $S \times \Delta$, $\epsilon \in \Delta$ marks the end of the stack, and $s_0 \in S$ is the starting state.[68]

Storage space of potentially infinite length enables PDSA to describe

---

[67] See Monod, *Chance and Necessity*, Chapter 4.
[68] See Birkhoff and Bartee, *Modern Applied Algebra*, p. 419.

behavior of some complexity. Such systems are not necessarily governed by an immediately prior set of states and symbols, so their talents are suited for mirroring sophisticated cellular processes involving feedback loops, controls of various sorts, delays, and the like. The bacterial cell, of course, suffers some idealization when set as a PDSA: bacterial systems quite obviously must complete their computations within close limits of time and space, and in any case specimens of *E. coli* hardly have boundlessly large memories at their disposal. Constructing an automaton that carries out lifelike activities is quite difficult. The von Neumann models are gargantuan, and they turn out to have the power of Turing machines. But the outlines of the modeling can be sketched in a tentative way.

Biological PDSA embody algorithms for the conversion of codons into strings of nucleotides. They thus control the nature of the proteins that are sequenced and the order in which they are synthesized. *States of the bacterial cell* can therefore be identified with *states of a PDSA*; the full set of *codons* formed from a nucleotide alphabet can be identified with its set of *input symbols*. That fixes X, S, and, with some minor fiddling, Δ. The elements ϵ and S_0 can be marked arbitrarily. The set of sequences the machine accepts when confronted with an empty stack comprise the complete set of nucleotide strands: a simple coding transformation takes them into proteins, so that for all practical purposes the output of a biological PDSA can be identified with the set of functioning proteins the machine computes. The function δ presumably has the capacity to specify an infinite class of objects (there would be nothing interesting in a machine that boiled itself down to a list), but varying uncertainties about the remaining details make the modeling quite artificial. Without more complete information about the transition-state functions, we are left with a class of machines that have fixed properties and limitations but no specific computational powers. Yet an automata-theoretic approach so fashioned satisfies some algorithmic intuitions about the cell: the feeling that the growth and regulation of the cellular machinery is basically a recursive process in which a finite set of elements is teased into a complex construction.

The problem of constraints and the space of polypeptides

The introduction of automata carries conceptual constraints, and such satisfactions as the model provides may be diluted when its liabilities are toted up. Formal systems require rules, and rules require representation. This involves a note of strain. Much more important is the problem of reconciling the supposedly stochastic processes by which information changes—this on the central dogma—and the insusceptibility of formal objects to random perturbations. Machines can be designed that incorporate probabilistic features: their transition-state functions push the automata to states embodying different probabilities, and the resulting machines are not deterministic. But the central dogma sets out something quite different: a system in which elements of the base vocabulary change at random and thus form output sequences wholly unlike any that the machine is prepared to compute. PDSA are close to Turing machines, and a Turing machine can be depicted as the program for an actual digital computer. Computers whose programs are fiddled with randomly, simply jam.

This might suggest the saving thought that biological PDSA instantiate an attitude of computational indiscriminacy, with *every* possible sequence included in the set of output sequences. But there is something conceptually unwholesome about this plan. Automata are algorithms, and every algorithm embodies a portion of the world's computable powers. Computability calls to mind procedures that are largely *mechanical*; but it also evokes the existence of *constraints* on the possibilities involved in converting inputs into outputs. The bacterial cell, with its exquisitely refined biochemical mechanisms, could hardly be under the control of an object inherently liable to a form of babbling.

This is purely an automata-theoretic point; Schutzenberger has made it vividly in the context of his lectures to an audience of uncomprehending biologists.[69] His argument, crudely put, which is the way he put it himself, is just that random changes destroy meaning in any sense of biological meaning, however loose. I think this is correct, but the essential idea may be recast more successfully in another way.

Although the earth seems to swarm with bizarrely constructed creatures, the stock of proteins that go into their construction must cut a narrow swath through the assortment that is combinatorially available.[70]

Proteins that have actually been strung together in a working cell share the property of *functional usefulness*. In one fashion or another they have been integrated into the cell's internal economy. A protein is *viable*, let us say, just in this case. Viability has little to do with the protein's contribution to cellular survival: most proteins that go into the grand stock of 10^{50} proteins sequenced since time immemorial are viable, but thousands among them must have contributed their share

[69] M. Schutzenberger, "Algorithms and the Neo-Darwinian Theory of Evolution," in P. S. Moorhead and M. M. Kaplan, eds., *Mathematical Challenges to the Neo-Darwinian Theory of Evolution* (Philadelphia: Wistar Institute Press, 1967).
[70] See M. Eden, "Inadequacies of Neo-Darwinism as a Scientific Theory," in Moorhead and Kaplan, *Mathematical Challenges*, p. 7.

to the occurrence of cellular catastrophes. Neoplastic diseases of the cells that regulate the human immunodefense system, for example, are accompanied by disastrous changes in the serum proteins (*plasmacytic dycrasias*). Multiple myeloma, for example, is often marked by Bence Jones proteinuria. Instead of producing a complete immunoglobulin molecule, the diseased immunodefense system spins off large quantities of the Bence Jones protein. Renal damage follows, followed often by massive kidney failure and then death. But there is nothing "unviable" about the Bence Jones protein, at least not in the sense I am prepared to attach to the word.

Viability is a property much in evidence among existing proteins; whatever the vagaries of the ways in which viability is smeared across the entire space of polypeptides, most proteins actually on hand indisputably have achieved a sort of functional usefulness in some context or cell. Their organization, then, should tell us something about whatever it is that renders the viable proteins viable. It is plausible, following this line of thought, to think of the space of all proteins *topologically*, with an associated *metric $D(P_i, P_j)$* between protein strands defined in terms of protein *homology*. Two proteins are perfectly homologous just in case they are composed of the same residues arranged in the same order. In that case $D(P_i, P_j) = 0$. Pairs of proteins with no homologous stretches whatsoever can be pegged at an arbitrary distance, and D can easily be defined to accommodate the proteins that lie between the extremes. The topological space that results is perhaps degenerate, since physically distinct proteins at no distance from each other must be treated as if they were identical. It might be more natural to pass from the space of all possible proteins to a slightly more abstract space composed of equivalence classes formed, say, under the relationship of perfect homology. Homology itself may prove too restrictive a notion on which to base a protein metric; obscure properties of the polypeptides might be more intimately connected with cellular viability. No doubt the opportunities for misdirection are great, but there is something very natural about the thought that the mysteries of cellular usefulness are bound up with something as simple as the ways in which the amino acids are arranged.

The obvious question, then, for those who have reached this point intact, is whether the stock of existing proteins—a subset of that vast group of proteins which have so far seen sequencing within the cell—comprises a random sample of the space of all possible proteins. If so, it would be hard to resist the conclusion that virtually any protein might find functional usefulness: the nucleic acids achieve a bias toward viability simply because the viable proteins exhaust the space of polypeptides. On the other hand, the existing proteins might occupy a portion of the protein space whose members are all closer to each other than to the other possible proteins. Viability would have to be the property of a depressingly small number of polypeptides: vast regions of the space would simply be closed to life.

Murray Eden and others have argued that the available proteins are by and large closer to one another than they have any reasonable right to be. Alpha and beta chains of human hemoglobin A, for example, contain 140 and 146 residues, respectively: when arranged for optimal homology, they turn out to agree in 61 places, differ in 76, and admit 9 gaps. One chain was probably derived from the other, or both from a common precursor. The number of point mutations leading from alpha to beta hemoglobin seems to be somewhere in the neighborhood of 112. Yet,

if we look at the distribution of residues, they are quite similar, with a mean difference of about 1.5 per amino acid type. . . . One would not have anticipated that the distribution for the amino acids occurring in the alpha but not the beta chain is very close to the distribution of residues occurring in the beta but not the alpha chain.[71]

Some statistical results point the other way. In 1956, Gamow, Rich, and Ycas took a long look at a tiny batch of proteins and concluded, after some communion with the computer, that the residues were distributed at random: the amino acids, in fact, obeyed nothing more interesting than a simple Poisson distribution.[72]

In a later article, Sorm and Keil went over much the same evidence but came to just the opposite conclusions.[73] They reported themselves much impressed by the number of homologous proteins, obvious deviations from random in Cytochrome C, repetitions of peptide sequences in positions 78–85 of the hemoglobin alpha chain, and mysteriously symmetrical groupings of sequences in the heme protein of the photo-

---

[71] Ibid.

[72] G. Gamow, A. Rich, and M. Ycas, "The Problem of Information Transfer from the Nucleic Acids to Proteins," *Advances in Biological and Medical Physics* (April 23, 1956).

[73] F. Sorm and B. Keil, "Regularities in the Primary Structure of Proteins," *Advances in Protein Chemistry*, Vol. 17 (1962), p. 167. See also A. Kreywicki and P. P. Slonimski, "Formal Analysis of Protein in Sequences," *Journal of Theoretical Biology*, Vol. 17 (1967), pp. 137-158, and J. M. Zimmerman et al., "The Characterization of Amino Acid Sequences in Proteins by Statistical Methods," *Journal of Theoretical Biology*, Vol. 21 (1968), pp. 170-201.

anerobe *Chromatium* (strain D). Fox and Nakashima have also reported some evidence of unsuspected order in a model of a prebiotic polymer.[74]

The hard evidence thus has a certain undeniable wobble. In part, this is a matter of a small data base. Only fifteen or so proteins have been completely sequenced; less than 1000 have been sequenced altogether.[75] Then there is the possibility—touched on already—that the proteins embody bizarre strategies of organization. Still, there is enough in the way of evidence to suggest that the stock of proteins is small and significantly bunched together; the entire vast space of polypeptides does not swarm with proteins that are available for life. This implies, of course, that viability is a property in relatively short supply among the proteins. Sequences of amino acids are but coded images of strings of nucleotides; thus, much the same point can be made in the context of the nucleic acids, with the viable nucleotide sequences construed simply as those which code for viable proteins. The thinness in the spread of viable polypeptides reflects a winnowing already effected in the set of possible genetic messages. A bacterial cell contemplating genetic change must therefore make do with just a fragment of the combinatorially conceivable nucleotide strands: constraints on the number of molecular messages manage the process of changing biological information within fairly narrow confines.

Constraints on this order do not exist—or so one would suppose from a reading of the central dogma. New proteins arise via random changes among the nucleic acids. Nothing in such an arrangement suggests the sources for a tilt toward viability. Except for physical biases (hot spots), reregistration of the proteins occurs with no discernible preferences or revelations of order. And the ultimate physical determinants of change—radiation, unusual chemical activity, etc.—fall on the nucleic acids in a way that says nothing about the possibilities for genetic organization:

at the two extremes—the nucleotides and the proteins—is not even chaos out of which one might believe that a certain regularity might emerge . . . but two systems having structures which are *a priori* not more in agreement than in conflict.[76]

So much the worse, one is tempted to say, for the central dogma.

[74] S. W. Fox and T. Nakashima, "Indications of Order in a Model of Prebiotic-like Polymer," in Moorhead and Kaplan, *Mathematical Challenges*, p. 123.
[75] Details are set out in M. O. Dayhoff, ed., *The Atlas of Protein Sequence and Structure* (Baltimore, Md.: National Biomedical Research Foundation, 1969).
[76] Schutzenberger, "Algorithms and the Neo-Darwinian Theory of Evolution," p. 74.

If the possibilities for life fill out a small space, then something more interesting than a blind if hungry capacity to capitalize on genetic error *must* be at work in the generation of biological change. Some mechanisms will be purely physical. Certain linear sequences of proteins might be incapable of collapsing into globular configurations that the cell can use. Others might be code-theoretic. The majority of the proteins reflect the code's degeneracy; the code may be self-synchronizing.

Such constraints, however, illustrate no general principles that might serve to control admissible strings of nucleotides. The original invocation of automata as models for the bacterial cell carried with it as conceptual baggage the thought that biological automata not only arranged the affairs of the cell, but fixed in recursive form computational powers that were sufficient to segregate the viable from the unviable proteins. This was the imagined sticking point for biologists; modifying the central dogma merely to smooth the way for automata seemed an arrangement of altogether excessive zealousness. But now we are set to wondering about the mechanisms that constrain the code—this in response to a purely biological line of argument. Constraints evoke algorithms, algorithms automata; there is no reason now not to appreciate the hypothesis that purely syntactic restrictions on admissible sequences are set by biological machines.

What is needed to complete such enthusiasms is some system of insights into the specifications adopted by the bacterial cell. Biological automata must somehow be faithful to the diversity of life that is accessible from a particular species; at the same time, impossible options must be foreclosed. Linguists, in devising a generative grammar for a natural language, must also concoct machines that meet some of these demands, and the purely linguistic problem of producing a "natural form" for grammars is analogous precisely to the biological problem just discussed.[77] Biological PDSA modeling, say, the bacterial cell could emerge as machines that maneuver through only a limited repertoire of genetic reassemblies. Such devices would instantiate algorithms that fix in advance the strings of nucleotides that are accessible to a given organism: automata that register shifts or changes in nucleotide sequences would stand to each other as dialects of a language. There would be little trouble in preserving a sort of vestigial segment of the central dogma in such machines: mutations could continue to comprise a set of equally probable events just so long as they were confined by constraints fixed on admissible strings of codons. Changes resulting

[77] See David Lewis, "Languages, Language and Grammar," in G. Harman, ed., *On Noam Chomsky* (New York: Doubleday, 1974), p. 260, for details.

in strings that could not be generated would simply not arise or, failing that, would arise without effective genetic expression.

Automata of such sophistication would be marvelous machines—an algorithm that accomplishes a winnowing among the proteins instantiates a definition of life itself.[78] Unfortunately, the neo-Darwinian theory of evolution, such as it is, provides little that might be used to explain their origin. Enfeebled automata do not suddenly acquire great computational powers; the contrary hypothesis that life began with a set of cells fully committed to just the viable proteins has little save elegance to commend it. A similar problem arises in accounting for the origin of the first cells. Various chemicals in the prebiotic environment might have assembled themselves into amino acids: it is a great deal harder to imagine how a batch of amino acids managed to coagulate into a functioning cell.[79]

One line of research, begun by George Pieczenik, involves the hypothesis that symmetries of certain sorts may be universal in nucleotide sequences.[80] For example, in the R17 phage, one comes across palindromes of the following sort:

Protein:	ASN	SER	ARG	SER
Codon:	AAC	UCG	CGC	UCA.

Evidently their general form is $\varphi \chi \varphi^*$, where φ and φ^* are mirror images and χ is a point of symmetry. It is a configuration that occurs with some regularity in the informational macromolecules—proteins—as well as nucleic acids. PDSA can in fact generate infinitely many strings of this sort; and if *only* such strings are generated, their set is not regular, and one must employ at least a PDSA to compute them all. The ideal cell might be biased toward symmetries of a simple sort, with the actual

---

[78] L. J. Fogel has performed experiments on finite-state automata in which the automata mutate freely—via random changes in transition tables—and are then subject to selection. But, plainly, if the accessible portions of the polypeptide space are sparse, no random mechanism could ever turn them up. Some algorithmic direction will always be required; arbitrary random changes performed on generative grammars are unlikely to yield new human languages. See L. J. Fogel, M. J. Walsh, and A. J. Owens, *Artificial Intelligence Through Simulated Evolution* (New York: John Wiley & Sons, 1966).

[79] See H. Jacobson, "Information, Reproduction and the Origin of Life," *American Scientist* (1957), pp. 119-127, for an argument that from an information-theoretic viewpoint life is impossible.

[80] For details, see G. Pieczenik, P. Model, and H. Robertson, "Sequence and Symmetry in Ribosome Binding Sites of Bacteriophage f1 RNA," *Journal of Molecular Biology*, Vol. 90 (1974), pp. 101-214.

bacterial cell reflecting the disarray and arbitrariness in nucleotide order that results from purely physical limitations of space or stereochemical configuration. The thought that various symmetrical groups are connected by specific *transformations* also seems natural. Progress is likely to be a matter of putting together a set of insignificant algebraic results in the hope of stumbling on some general principles of genetic organization.

All this, of course, is speculative, even high-flown. The experimental evidence is meager; it is unlikely that many biochemists will rush to an area in which orthodox techniques are unhelpful. In any case, it hardly seems likely that the syntax of the nucleic acids will have a very natural biochemical explanation. Although the recursive properties of biological entities may all have a common character, there is no assurance that they will reflect anything interesting about the chemistry of the cell. The order embedded in the genetic code may not be chemical. A biological materialism, refined and made sophisticated, may prove an ironic doctrine.

Biological automata naturally model those processes that seem to characterize the bacterial cell; in this respect, the appeal to mathematical systems theory is largely immune to the standard criticisms of mathematical infelicity. Although biological automata may well be able to *express* the properties exhibited by the bacterial cell, however, there is nothing to indicate that automata theory will be able to *explain* why, for example, the space of polypeptides accessible to life is just the space that it happens to be. Nor is it at all clear why the structure of syntax at the molecular biological level should, so far as we can judge, reappear at the gross level of the full-fledged organism ultimately created out of instructions stored in nucleic acids.[81] Of course, animals that are closely related to one another simply as species turn out to have sequences of nucleic acids that are close simply as ordered linear strings. This is what we expect, but hardly what we can explain using

---

[81]This is the problem of *correspondence*, which, as far as I know, only Schutzenberger has discussed explicitly (see "Algorithms and the Neo-Darwinian Theory of Evolution"). I have discussed the problem raised by what seems to be an unexplained correspondence between gross and molecular levels before an interdisciplinary colloquium at UCLA, and I was surprised by the insistence of so many biologists that what I have dubbed a problem is nothing of the sort. So perhaps I can recast the issue again. What we have, to begin with, are two *different* metrics defined between sets of nucleic acids on the one hand, and between sets of organisms on the other. It is the second metric to which we appeal when we say, for example, that two dogs are closer to each other than either is to a moose; the first to which we appeal in saying that strands of nucleotides differing by one residue are closer to each other than either is to a completely different strand.

concepts only as strong as those found in automata theory or the equivalent theory of formal grammar. From this it follows that insofar as automata-theoretic models form images of the powers and properties of the bacterial cell, they do not model the whole of those properties and powers and, in fact, say nothing of interest about the most fundamental properties of all.

And in this respect, the application of automata theory to biological phenomena is typical of systems theory generally.

Now consider the following diagram, labeled in an obvious fashion:

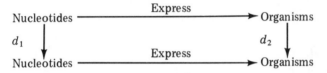

Abusing mathematical language slightly, one might say that this diagram "commutes" in the following sense: distances d_2 measured along the right-hand arrow are inevitably proportional to distances d_1 measured along the left-hand arrow. As far as I can see, there is no reason whatsoever why this should be the case. For a recent discussion involving this issue, see M. King and A. C. Wilson, "Evolution at Two Levels in Humans and Chimpanzees," *Science*, Vol. 188 (April 11, 1975), pp. 107–116; the authors do not recognize in its full generality the significance of the issues they raise.

Epilogue

Only the absolute idealist and the student who has seen Krishna think of the Universe-as-a-Whole in the sense that there is nothing but the Whole. Others subdivide, resting easy finally with a motley that includes stars in the staring sky as well as those coiled heaps of nucleic acids that biologists assure us give form and content to tons and tons of blubbery protoplasm.

The mind recoils, though, at the thought of a universe incoherent as a heap. In physics it is only at the level of general relativity that there are vast integrative schemes and a satisfying sense that an act of the intellect has at last proposed something suitable to the sweep and heft of the observable world itself. To a man trained in the techniques of geology, ceramic engineering, or international trade, the sparse usufructs of his daily toil may seem dissective, partial, and incomplete.

Systems analysis is very much a modern movement, inseparable really from the raucous growth of the new intellectual technology of game theory, decision theory, cybernetics, information theory, and mathematical systems theory. Nor do the various separate sections of the theory, judging now strictly from the perspectives of the preceding chapters, really have very much more in common than their name. Still, it is possible to see in outline the lineaments of a system of intellectual desires imperfectly realized by the various virtuosos of the systems-analytic arts. There is in systems analysis an inexpungible craving for generality, and it is this craving that makes common the concerns of the various authors I have trounced.

The craving for generality is, of course, an immemorial intellectual inclination; without it, mathematics would be sterile and physics rather uninteresting. What separates systems analysis from such serious studies is simply the gap that inexorably opens between the conception and execution of a set of intellectual ambitions. So much of systems analysis is mathematically meretricious. But even here there is a continuum of achievement. General systems theory marks what Scientologists, I believe, call a point of minimal adequacy. This is a movement that is all craving without content. One goes up from there: in the vast schemes of *World Dynamics* and *The Limits to Growth*, there is the *essential* systems analysis, distilled as it were, and just successful enough scientifically to avoid being destroyed outright by a bark of contempt. Here the elements are all perfectly arranged: vast hopes, pressing social needs, and a theory that has in ordinary differential equations objects of perfect mathematical propriety. Steadily inflated by an almost fabulously daring vulgarity, the theory bursts in the end simply because its authors see something disagreeably unscientific and certainly unprofitable in remarking merely that since human population growth is sure to have limits, something rather unattractive is bound to happen when those limits are reached. Mathematical systems theory is—like the theory of ordinary differential equations from which it evolved—a perfectly serious mathematical discipline. I have split the field into separate halves: the theory of linear systems and automata theory. It is only when these subjects are *appropriated* for analyses executed in the political or social or biological sciences that one has systems analysis, with results that are generally, although not inevitably, disappointing. Gassy efforts such as Easton's are objectionable on a score of points, but there is a certain redeeming inflation of content for theories that trade on traditions in engineering or mathematical systems theory. And in biological systems analysis, I think, there is a kind of successful matching of ambitions to available mathematical models that is entirely rare elsewhere in the systems-analytic arts. But even here, that liberating sense of having understood a phenomenon such as the bacterial cell *generally*, which does result from an automata-theoretic perspective, does not yet by any means afford understanding that is in any sense *complete*. Even if we grant that the cell is an automaton, it is clear at once that the full explanation for its powers and properties must go beyond automata theory entirely. In this respect, I shall guess, the history of linguistics will be repeated in the development of molecular biology.

There may be a moral in this record of grand efforts brought low by insufficient means, some cagey bit of philosophical wisdom. But aside from the obvious counsel that in great things great ambitions without great theories are insufficient, I do not know what it is.

Index